Impact!

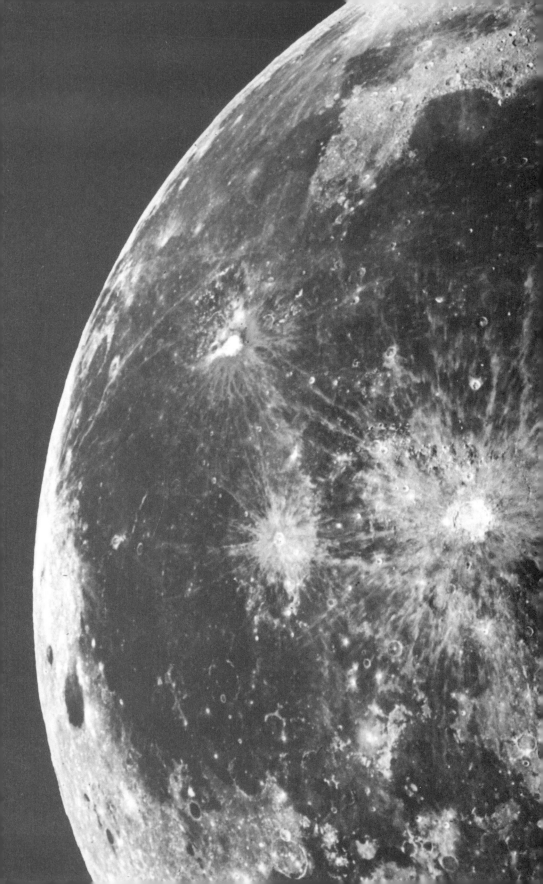

Impact!

The Threat of Comets and Asteroids

GERRIT L. VERSCHUUR

OXFORD UNIVERSITY PRESS
New York   Oxford

Oxford University Press

Oxford   New York
Athens   Auckland   Bangkok   Bogotá   Bombay
Buenos Aires   Calcutta   Cape Town   Dar es Salaam
Delhi   Florence   Hong Kong   Istanbul   Karachi
Kuala Lumpur   Madras   Madrid   Melbourne
Mexico City   Nairobi   Paris   Singapore
Taipei   Tokyo   Toronto   Warsaw

and associated companies in
Berlin   Ibadan

Copyright © 1996 by Gerrit L. Verschuur

First published by Oxford University Press, Inc., 1996

First published as an Oxford University Press paperback, 1997

Oxford is a registered trademark of Oxford University Press

Library of Congress Cataloging-in-Publication Data
Verschuur, Gerrit L., 1937–
Impact!: the threat of comets and asteroids /
Gerrit L. Verschuur.
p. cm.   Includes bibliographical references and index.
ISBN 0-19-510105-7
ISBN 0-19-511919-3 (Pbk.)
1. Comets. 2. Asteroids. 3. Impact. 4. Earth. 5. Evolution
(Biology) I. Title.
QB721.V48   1996
551.3'97—dc20   95-45030

35798642

Printed in the United States of America
on acid-free paper

PREFACE

The discovery that a comet impact triggered the disappearance of the dinosaurs as well as more than half the species that lived 65 million years ago may have been the most significant scientific breakthrough of the twentieth century. Brilliant detective work on the part of hundreds of scientists in analyzing clues extracted from the fossil record and geological strata, and from a census of near-earth space, has allowed that mass-extinction mystery to be solved. As a result we have gained new insight into the nature of life on earth.

Impact! The Threat of Comets and Asteroids offers an overview of what has been learned about the nature of cosmic collisions and the remarkable twist the new knowledge has given to the story of our origins. Comet and asteroid impacts may be the most important driving force behind evolutionary change on the planet. Originally, such objects smashed into one another to build the earth 4.5 billion years ago, after which further comet impacts brought the water of the oceans and the organic molecules needed for life. Ever since then, impacts have continued to punctuate the story of evolution. On many occasions, comets slammed into the earth with such violence that they nearly precipitated the extinction of all life. In the aftermath of each catastrophe, new species emerged to take the place of those that had been wiped out.

Recognition of the fundamental role of both comet and asteroid collisions in shaping evolutionary change means that the notion of survival of the fittest may have to be reconsidered. Survivors of essentially random impact catastrophes—cosmic accidents—were those creatures who just happened to be "lucky" enough to find themselves alive after the dust settled. No matter how well a creature may have been able to survive in a particular environment before the event, being thumped on the head by a large object from space is not conducive to a long and happy existence.

The evidence that our planet continues to be vulnerable to devastating bombardment from space is overwhelming. When this possibility was again brought into focus, after having been out of fashion for most of this century, many scientists and a large fraction of the public thought the idea ludicrous. A major newspaper even coined the expression "giggle factor" in reference to the reaction of those who couldn't take the idea of asteroid impact seriously. That expression is no longer heard. Some critics conjured up visions of the storybook character Chicken Little running about cackling "The sky is falling, the sky is falling!" They, too, have been stilled.

Our new understanding of why the dinosaurs and so many of their contemporary species became extinct has revealed the earth as a planet not specifically designed for our well-being and one that continues to be the target of comets as well as asteroids. From time to time, life is rudely interrupted by shattering events on a scale we can barely imagine.

The reality of the threat of comet impacts was brought home to everyone in July 1994 when fragments of a comet slammed into Jupiter's atmosphere to produce a stunning series of explosions that were seen from earth. Fortunately, we watched from a safe distance. If anything remotely similar had happened here, few human beings would have been left to think further about comets, asteroids, or anything else for that matter.

For more than two centuries the possibility that the earth might be struck by comets has been debated and three questions have been raised from the start: will a comet again hit the earth and, if so, when; might comet impact lead to the extinction of species; and is it possible that the flood legends from so many cultures around the world could be accounted for by past comet impact in the oceans that triggered enormous tsunamis? Within the last two decades of the twentieth century, affirmative responses to the first two questions have become a part of the scientific mainstream. It is remarkable to see how rapidly scientists who were once skeptical that impacts might again occur, or that they might be blamed for the extinction of species, now argue about the details of the mechanisms involved. Many of them have also begun to think seriously about the fate of civilization in the face of the threat of comet and asteroid impacts.

The third question has begun to experience its own revival, but here the implications of an affirmative answer reach beyond the scientific. Great prejudice exists both for and against the idea that the Deluge (the biblical flood) was a real event triggered by asteroid or comet impact. To allow this possibility impinges on the long-held beliefs of many people who would prefer the event to have more metaphysical overtones. However, in the light of the breakthroughs that have been made in understanding the nature of cosmic collisions, new light is being cast on what may lie behind ancient legends, sagas, and myths that tell of terrible floods that once ravaged the world.

In comparison with more immediate threats to the continued survival of our species (acid rain, destruction of stratospheric ozone, the greenhouse effect, and overpopulation), the danger of comet or asteroid impacts may seem remote. The problem with impact events, however, is that their consequences are so awesome that we can barely imagine what it would be like to be struck by a large object from space. And there would be limited opportunity for reflection following such an event.

There is another side to the threat of comets and asteroids. Cosmic collisions clearly set the scene for the emergence of *Homo sapiens,* a species that recently became conscious enough to design and manipulate instruments that allow it to explore beyond its senses. In so doing, our species came to behold how it fits into the cosmic scheme of things.

Once we appreciate that impact catastrophes have shaped life as we know it, both over the long term as defined by mass extinction events and in the short term as identified in historical data, and that such events will happen again in the future, how will this awareness alter the way we look at ourselves in the cosmic context? Will we let nature take its course and trust to luck that our species will survive the next violent collision? Or will we confront and deal with the difficult behavior decisions that may yet influence the destiny of all life on earth? More specifically, the manner in which we act on the insights that have been garnered about collision catastrophes will almost certainly determine whether or not civilization has a long-term future.

Many details referred to in our story are still controversial. Strong opinions exist on opposite sides of all three of the questions listed earlier. Debate surrounds the details of extinction events as well as virtually every aspect of the nature of comets and asteroids. Debate is particularly heated as regards the role of impacts in directing the course of human history. All of this is very exciting. The subject of impacts and their relationship to mass extinction, as well as the implications for the future of life on earth, is in a state of ferment, a symptom that something significant is brewing. Just how significant every reader will have to judge after reading this book.

These then are some of the topics explored in *Impact! The Threat of Comets and Asteroids.* Ultimately we must ask whether we find the risk of future impact to be sufficiently great to merit doing something to avoid it. Many dangers posed by living in a modern technological society are far more likely to cost us our lives, but that is not the point. Rare comet or asteroid impacts may cost <u>all</u> of us our lives. So how will the threat of comets and asteroids fit into our thinking? We can only answer this question after we have learned a great deal more about the nature of the danger.

Lakeland, Tenn.   G. L. V.
March 1995

ACKNOWLEDGMENTS

Work on this book has benefited tremendously from the input and help of many friends and colleagues as well as people whom I have never met who offered their input and advice. My interest in the subject was sparked as long ago as 1975 when I gave a series of lectures at Naropa Institute in Boulder, Colorado, on the cosmic unity one becomes aware of when one appreciates that life on earth does not exist in isolation from the rest of the universe. In fact, the nature of terrestrial evolution is intimately tied to many phenomena acting out on a cosmic scale, including starbirth and stardeath and the motion of the sun and solar system through the Milky Way. The awareness that grew out of those lectures led to my writing *Cosmic Catastrophes* (Reading, Pa.: Addison Wesley, 1978).

Since then the subject of the relevance of the cosmic environment to the nature of life and evolution has received a stunning boost from the discovery that collisions with comets and asteroids have mediated terrestrial evolution in dramatic ways in the past, and that they will do so again in the future.

Some of my colleagues whom I wish to thank for helping me better understand the phenomena described in this book include Victor Clube, Bill Napier, Duncan Steel, Brian Marsden, Mike A'Hearn, David Morrison, Tom Gehrels, Robert Jedicke, and Clark Chapman. Also, Anne Raugh's help in allowing me to enjoy the Jupiter impacts via the Internet is much appreciated.

A special thanks are due to Von del Chamberlain of the Hansen Planetarium in Salt Lake City for the opportunity to collaborate on a script on the subject of cosmic catastrophes in 1991, which rekindled my interest in the topic.

Three key players, Alan Hildebrand, Glen Penfield, and Robert Baltosser, helped me unravel the essence of the discovery story told in chapter 2.

For help in providing illustrations, I wish to thank Richard Dreiser, Glen Izett, Don Yeomans, David Malin, Kathy Nix, Wendy Wolbach, Alan Hildebrand, Glen Penfield, Richard Grieve, Janice Smith, Steve Ostro, Richard West, H. U. Keller, Syuzo Isobe, Robert Jedicke, and Roy Foppiano, who showed me some wonderful old books that considered the threat of impact.

Robert Jedicke acted as a gracious and informative host during my visit to the Spacewatch telescope on Kitt Peak near Tucson.

I also wish to express my appreciation to Jeff Greenwald for sending back reports of his global travels, to Leroy Ellenberger for producing fascinating background material on catastrophism and leading me to making further contacts as my research into the subject proceeded, to Alexander Tollmann for sharing his ideas, and to Herbert Shaw for a friendship that developed out of our discussion of the nature of terrestrial impacts.

I am particularly grateful to Tom Mitchell as well as an anonymous reviewer for valuable and crucial suggestions that helped to improve this book.

Finally, I am grateful for the unstinting support of my wife, Joan Schmelz, during the years that this book marinated, when no publisher seemed convinced that the subject was worthy of a detailed story, a situation that changed dramatically in July 1994 when Jupiter suffered the humiliation of being slapped by twenty-one comet fragments in front of everyone on earth. To those who shared their observations of that event online, thank you.

CONTENTS

Impact!

1

THE KILLER STRIKES

AT the beginning of the nineteenth century, French paleontologist Baron Georges Cuvier recognized that many fossils represented the remains of species that no longer roamed the earth but were only to be found in certain rock strata. To convey what this discovery meant, he painted a vivid picture. "Life on earth," he wrote, "has been frequently interrupted by frightful events." A modern commentator, Derek Ager, likened the tale revealed by the fossil record to that of the life of a soldier: "Long periods of boredom and short periods of terror."

The periods of "boredom" are what we experience for most of our lives, when all is well with the world. That is the way we like it. The climate is benign and the seasons come and go in an endless and reassuringly predictable procession, and we survive nicely without being threatened by nature. Sometimes the spell is broken by a catastrophe; a tornado ripping at our house, a flood washing it away, or fire engulfing all in its path. After the terror has passed, all is peaceful again.

Sixty-five million years ago a catastrophe of awesome proportions struck our planet. Something happened to wipe out the dinosaurs as well as about 60 percent of all species that lived at the time. After a century or more of sifting through incriminating evidence left at the scene of the crime, scientists have at last identified a comet or asteroid colliding with earth as the killer. The victims of this headlong collision on earth were felled by the devastating explosion of impact or killed in its aftermath.

The comet impact of 65 million years ago was not the first of its kind, nor will it be the last. To appreciate how serious the danger is, let's start with the dinosaurs. Their history has slowly and laboriously been pried from ancient rock and clay layers in which fossils are preserved.

Fossilization happens to plants and animals whose remains sink into the muddy sediments at the bottom of oceans, seas, lakes, rivers, and streams where they are preserved in what will later become layers of rock. Preservation occurs when the tiny spaces in bone, for example, become so impregnated with minerals in the water that the bone itself takes on a rocklike hardness. This process is very slow, but if the animal carcass is promptly covered by sediments, mud, or even volcanic ash, the structure of bones or shells is protected from further damage and becomes perfectly preserved.

Without nature's skill at fossilizing animals and plants, we would have no information whatsoever about the fauna and flora that existed a long time ago. Fortunately, nature has been kind and paleontologists have developed the skill to read the fossil story in considerable detail. They have gathered a lot of information about creatures that died natural deaths, and even more about what happened at times when entire species were wiped out. It was his awareness of the fact of mass extinctions that caused Cuvier to conjure up his image of terror.

William Smith, a civil engineer and builder of canals in England around the beginning of the nineteenth century, was the first to realize how to estimate the age of fossils. He noticed that different types of fossils were found in different rock layers and that a certain fossil type found in one layer was never seen in another. Furthermore, a particular variety was found only once in the cross sections of rock he studied as part of his canal building ventures. Under the assumption that deeper layers of rock were older, a picture began to emerge in his mind of a possible sequence of fossil ages.

Stratigraphy, the science of estimating relative ages of rock layers, involves making estimates of which fossil layer is older than another and relating fossils in one part of the world to those found elsewhere in similar rock structures. The actual age of a rock layer and associated fossils is now derived from radioactive dating, which, in the case of ancient rocks, is based on knowledge of the way certain radioactive elements decay. For example, uranium-238 decays to helium-4 and lead-206 (where the numbers refer to atomic weight) at a specific rate. In a time known as a half-life, half the atoms of uranium-238 experience this change. Based on measurements of the relative amounts of uranium-238 and lead-206 in a rock, it is possible to calculate the age of the rock. The point is that the record of the radioactive decay only becomes preserved from the time the rock actually hardened, whether from clay, mud, or lava.

Stratigraphic research combined with radioactive age estimates, many of

TABLE 1–1 Geological categories since the Precambrian

Era	Period	Epoch	Years ago
Cenozoic	Quaternary	Holocene (recent) Pleistocene	
			10 million
	Tertiary	Pliocene Miocene Oligocene Eocene Paleocene	
			65 million
Mezozoic	Cretaceous Jurassic Triassic		
			225 million
Permian			
Palleozoic	Carboniferous (Pennsylvanian and Mississippian) Devonian Silurian Ordovician Cambrian		
			570 million
Precambrian			
Origin of earth			4.5 billion

which involve other decay processes than the one mentioned here, has created a web of knowledge about past events that triggered fossilization on a global scale. The geological time series that emerged from these studies is illustrated in Table 1-1. Geologists name the rock layers according to where they are first noticed. This explains the origin of a label such as Cambrian, which refers to a rock layer found in Wales, or Cumbria, which was deposited about 570 million years ago.

The Precambrian *era* covers earth history from its formation, about 4.5 billion years ago, until the Cambrian, at which time a great diversification of life occurred and fossils began to be laid down in large numbers. It was then that many primitive life forms also began to develop hard shells,which were more readily fossilized. The eras following the Precambrian mark long periods during which the characteristics of life changed slowly. The era names in Table 1-1 refer to ancient life (Paleozoic), middle life (Mesozoic), and recent life (Cenozoic). Those eras encompass gross changes in the type of creatures that roamed our planet. *Periods* represent a further division into well-defined fossil structures that relate to characteristics seen within the eras. In the recent Cenozoic, periods are further subdivided into *epochs*. More recently, the fossil record became so rich and varied that far more detail can be recognized.

Distinctions in the types and numbers of fossil species reveal that on many occasions so many species went extinct at the same time that those have been labeled mass extinction events. They are so clearly evident in the fossil record that they are used to mark the boundaries between epochs and periods. The nature and cause of such events have been argued by paleontologists and geologists for more than a century, in particular as to whether the extinctions were gradual or catastrophic. About 150 years ago the gradualists won the argument over the catastrophists, for no very good reason as far as I have been able to ascertain. Many learned papers have been written about the debate but I can't help feeling that the ones I have read avoid a basic issue, one that is difficult for objective scientists to confront. On a personal level, the catastrophic point of view is psychologically abhorrent.

Ever since gradualism's victory over catastrophism, it has exerted a powerful bias on thinking about extinction, at least until a decade or so ago. Now the role of catastrophic events has again been brought to the fore by the recognition that the best-studied extinction of 65 million years ago, called the K/T boundary event, was caused by a comet collision with the earth. (K/T, pronounced as "kay-tee," refers to the boundary between the Cretaceous and Tertiary periods, Table 1-1, which one might expect to be called the C/T boundary. However, the initial letter, K, comes from the German version of the word Cretaceous, *Kreide*.)

It is nearly impossible to admit that our lives might be wiped out through a whim of nature, a chance event, and that may be why it has been so difficult to allow that dramatic mass extinctions of the past happened suddenly. It is even more difficult to admit that another such catastrophe might be triggered by a random collision between the earth and an object from space. This is the unpleasant likelihood suggested by the data. In a long tradition of separating themselves from their experiments, and any data or implications that emerge from those experiments, scientists have consistently presented a picture of mass extinctions that has become subtly distorted in the imagination. We assume that mass extinctions happened long ago and that nothing similar will happen again. This scenario is false. Not only might mass extinction events befall our planet again, but one is currently in progress. The most extensive mass extinction since the disappearance of the dinosaurs has been triggered by the impact of modern human beings on nature, but that is a story beyond the scope of this book.

The fossil record is known to contain evidence for at least nine major and a few dozen minor mass extinction events during the last half billion years. The best known of these, the K/T event, occurred 65 million years ago and saw the disappearance of about 60 percent of all species including as much as 75 percent of marine life, although the precise value of these estimates is still argued. For our story it does not matter whether the fraction was 40 or even 80 percent. What matters is that the mass killing was so great that the fossil record clearly shows a discontinuity in the numbers and types of fossils that were laid down around that time. Most significant, though, is that the dinosaurs went extinct at the K/T boundary, a fortuitous circumstance for us.

The full meaning of mass extinction events strikes more deeply when it is realized that the dinosaur extinction was not the greatest calamity ever. That horror belongs to the end-Permian catastrophe that occurred 245 million years ago when over 90 percent of all species were snuffed out in a single swoop, and life on earth very nearly came to an end. Words can barely begin to describe the enormity of such a catastrophe. No localized flicker of nature can account for the sudden demise of so many species at the same time. It required a global phenomenon of staggering proportions.

With the enormous interest created in recent years by the recognition that the dinosaur demise was the result of a comet impact, more and more evidence has been uncovered to show that species extinctions were in fact sudden. Dinosaur remains, for example, attest to their having flourished until the time of the mass extinction. Then, in a cosmic moment, the course of evolution was changed. Whether that moment involved months or years is impossible to discern from the fossil story alone, and in any case both are mere instants of geologic time. Some of the scientists now working to simulate the consequences of an impact explosion using computers think that the dinosaur extinction may have occurred within hours.

Between the fossil record, geological data, and astronomical observations of near-earth space, the clues needed to solve that mass extinction puzzle are at hand. As with any good detective story, the solution to a mystery is not always easy to come by, even when the clues are spread before you. Scientists require time to think about the meaning of their data as they look for patterns that will reveal what took place, and how and when it happened. They need to exercise the use of their mind's eye to consider how the "crime" was committed. Use of your mind's eye allows you to travel through space and time in imagination to get a feel for what it is that the geological, biological, or astronomical data reveal about the nature of nature. The closer you look at the data, the more clearly you learn to see. That is what geologists and paleontologists have been doing patiently for decades, if not centuries. Many individuals have added pieces to the puzzle and the overall picture now takes shape before us.

The dinosaurs were around for over 150 million years and, if they had not been wiped out, mammals would not have risen to dominate the world in their stead. After the dinosaurs were ushered off the terrestrial stage, the scene was set to allow mammals to diversify until, 65 million years later, one of their kind, *Homo sapiens,* rose to prominence. Our species recently evolved to become conscious and clever enough to invent agriculture, technology, and science, and we have used our newly developed mental skills to uncover the secrets of nature that carry the clues to our origins, and to our future. To put this another way, if the comet that triggered the K/T event had arrived 20 minutes earlier or later it

Figure 1-1 Photograph of the Cretaceous-Tertiary boundary rock layers at the Madrid Road site west of Trinidad, Colorado. Shocked quartz grains and an iridium excess are found in the centimeter-thick layer seen across the center of the photograph. (Courtesy Glen A. Izett, U.S. Geological Survey, Emeritus)

would have missed the planet and we would not be here now, talking, reading, or writing about any of this.

Now let's look at the evidence that a comet impact triggered the K/T extinction event. The clue that spurred so much study of this issue in recent years was found after painstaking work in analyzing the chemical composition of the clay layer that marks the K/T boundary (Figure 1-1). In 1980, the physical chemist Luis Alvarez at the University of California, Berkeley, and his colleagues (Walter Alvarez, Frank Asaro, and Helen Michel) announced the discovery of a thousand times more iridium in the K/T clay found in Gubbio, Italy, than is usual on the earth's surface. This may have been the single most important scientific breakthrough of the century, especially if one judges relevance on the basis of how a discovery affects our subsequent understanding of ourselves in the cosmic context.

What does iridium have to do with this? Iridium is a metal that is rare on the surface of the earth because it has long since disappeared to the center of the planet. Iridium is known as a siderophile, which means "iron loving." It sticks to iron and goes where the iron goes. When the planet was still molten in its formative years, the iron sank to the core, dragging most of the iridium with it. But then, you may ask, how do we know what is normal as regards how much iridium is expected on earth? The answer is found in astronomy. Observations of starlight and sunlight tell us about the average amount (or abundance) of various elements in the universe. The cosmic abundances are, in turn, accounted for by the action of nuclear processes in the hearts of stars where the elements are cooked up through fusion reactions. (Elements heavier than iron are brewed under even more dramatic circumstances, during explosions that mark the death of massive stars, explosions called supernovae.)

Data on cosmic abundances also come from the study of meteorites, rocky or metallic objects from space that fall to earth from time to time (see chapter 3). In particular, meteorites are thought to be relatively pristine examples of the matter from which planets were formed. Meteorite abundance data confirm that there is a certain fraction of iridium in the universe, a fraction considerably greater than found in rocks and soil at the earth's surface. Sometimes slightly elevated amounts of iridium may come surging out of volcanoes that tap deep into the earth's crust, but overall it is rare.

When Alvarez and colleagues discovered the excess iridium in the K/T clay, they realized that something special must have happened to cause the stuff to be deposited (Figure 1-2). So where had it come from? It must have come from space, they argued. The impact of a comet or asteroid could account for the iridium excess.

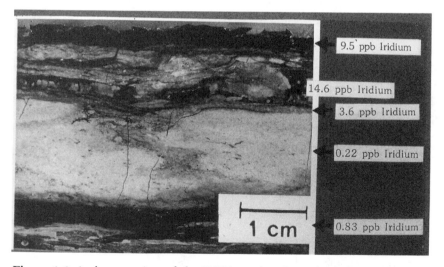

Figure 1-2 A close-up view of the K/T boundary layer from the Clear Creek North site, a few kilometers south of Trinidad, Colorado. In ascending order, the boundary interval consists of (1) carbonaceous shale of Cretaceous age, (2) a layer of white clay about 1.5 centimeters thick called the K/T boundary clay bed, or the ejecta layer, and marked with iridium content of 0.22 parts per billion (ppb), (3) a yellow iron sulfate clay layer about 0.5 centimeters thick containing shocked quartz and the peak iridium anomaly (14.6 ppb), also called the fireball layer, and (4) a thin Tertiary coal bed about 4 centimeters thick that would extend out of the top of the picture. (Courtesy Glen A. Izett, U.S. Geological Survey, Emeritus)

At this point we make a brief detour to explain the problem of whether the impactor was a comet or an asteroid. Comets (chapter 4) are thought to be mostly icy objects built around cores of dust mixed with ice, although rocky interiors cannot be ruled out. Asteroids (chapter 3) on the other hand are more rocky or metallic without being mixed with water ice. The distinction between these two types of objects has become increasingly blurred in recent years, as we shall see. The evidence that has recently come to hand suggests that the K/T impactor was a comet rather than an asteroid.

Whatever the details of the shattering impact, the resulting explosion blanketed the world in a dust layer that contained a thousand times more iridium than is normally found on the planet's surface. In order to account for the total amount of iridium deposited, the rocky part of the impactor must have been about 10 kilometers across with a mass of a trillion tons. Its head-on collision generated an explosion with the energy equivalent of a 100-million-megaton bomb. (This "yield" is measured with respect to the energy produced by a ton of TNT, with a megaton signifying a million tons.) In comparison, the explosive force of the Hiroshima atomic bomb was 13,000 tons (or 13 kilotons), while the largest nuclear weapons are around 50 megatons. That means that the K/T impact explosion was something like 20 million times more powerful than the most deadly hydrogen bombs ever detonated.

The blast of the K/T impact launched a staggering 100 trillion tons of material, vaporized cometary stuff plus terrestrial rock, into the atmosphere to encircle the globe, which explains why the K/T iridium signature has been found in more than 50 locations all around the world, in the so-called fireball layer.

In 1984 another link between a violent event and the cause of the K/T extinction was forged by Bruce Bohor and colleagues at the U.S. Geological Survey in Denver. They found shocked quartz crystals in the K/T layer. Evidence for the shocks is seen under high magnification where rows of atoms in quartz crystals have been disturbed (Figure 1-3). This can only be explained if the crystals were subject to great pressure, such as is associated with an impact or an explosion.

In 1985, Wendy Wolbach from the University of Chicago and her collaborators discovered that the K/T boundary layer also contains large amounts of soot produced by burning coniferous forests (Figure 1-4). The nature of the soot deposited in Europe as well as New Zealand is so similar that it suggests mixing caused by a global fire that began before the ejecta from the impact had even settled. It turns out that the global fires were ignited by the fireball produced at impact and helped along by a shower of flaming debris that fell over wide regions of the planet. The likelihood of a comet impact creating such havoc was confirmed by the remarkable events of July 1994 when 21 comet fragments slammed into Jupiter (chapter 14). The explosive plumes from some of those impacts were

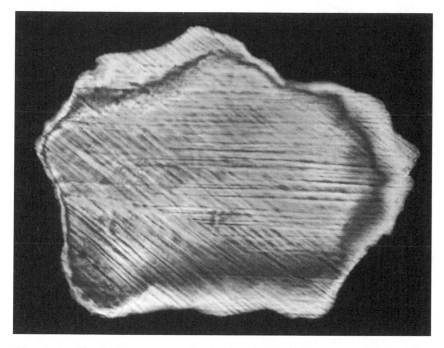

Figure 1-3 Shocked quartz associated with the K/T impact event found in the Madrid Road K/T boundary site, west of Trinidad, Colorado. Two prominent sets of shock lines are present. Such features have only been found at known terrestrial impact or nuclear explosion sites. The grain is 0.25 millimeters long. (Courtesy Glen A. Izett, U.S. Geological Survey, Emeritus)

so huge that, had they occurred here, the entire earth would have been showered with debris heated to incandescence as it reentered the atmosphere.

Following the impact of 65 million years ago, the sheath of soot, dust, and ash that enveloped the earth turned day into darkest night, terminating photosynthesis entirely, and plunging the planet into a deep freeze. Plants died and animals starved. The darkness lasted for months. Many survivors of the initial impact and its immediate aftermath froze to death in what has been called impact winter. Only those that could burrow for shelter and food had a fair chance of surviving the catastrophe. Creatures that did so included small, ferret-like mammals, which for tens of millions of years had managed to keep out of the way of tyrannosaurus and its ilk by, for example, living underground. After the dinosaurs were all dead, these mammals were well placed to flourish. During the subsequent millennia they diversified to fill the ecological niches opened as a result of the almost total cleansing of the planet. That is how it came to pass that 65 million years later one species of mammal, *Homo sapiens,* emerged to become the most powerful creature on earth. But something has not changed. Terrestrial life—all of it—remains vulnerable to devastating blows from space, at least until

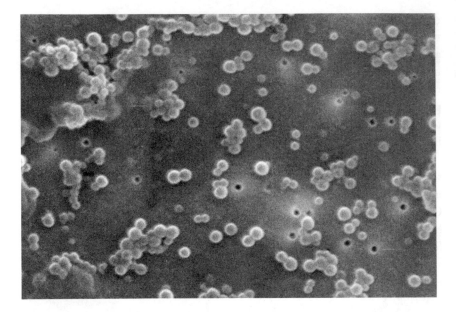

Figure 1-4 Soot particles from wildfires ignited after the K/T impact. These were found at the K/T boundary site at Woodside Creek in New Zealand. This micrograph has been enlarged 20,000 times. Soot particles cluster while the black disks are holes in the filter paper on which the soot was mounted. (Courtesy Wendy S. Wolbach)

such time as we decide to do something to protect our planet from a potentially lethal impact.

For a while after the discovery of iridium in the K/T boundary clay, arguments raged between proponents of impact and those who thought that heightened volcanic activity could have triggered the extinction events. For example, climate change triggered by volcanism might have driven species to extinction and the iridium could, perhaps, have been pumped into the atmosphere by volcanic explosions. This is possible if volcanoes tap deeply enough into the earth's crust to reach the region where the iridium is stored, but that is now regarded as all but impossible. Very few diehards still cling to the notion that volcanism was the primary trigger for the K/T extinctions.

The idea of terrestrial events being driven by rogue comets or asteroids slamming into the earth at first seemed like science fiction to many volcanologists and paleontologists. They denied vociferously that astronomically related causes could be a factor. How the argument proceeded has been documented by William Glen in his book *The Mass Extinction Debates: How Science Works in a Crisis.* There he demonstrates how the new idea struggled to take hold, which is, of course, a common phenomenon. Great insights seldom catch on like wildfire, whether in science, philosophy, politics, or virtually any other realm of human activity. Science, in particular, is too conservative a discipline to accept a new notion without a great deal of careful consideration. That is how it should be. As regards the impact theory for extinction, however, the idea turned out to be too powerful to resist. Within a year of being mooted in the context of iridium data to back up the theory, the idea spread quickly and captured the public imagination.

In 1989 another discovery helped sway opinion in favor of the K/T impact scenario. It was the detection of amino acids in the clay just above and below the iridium layer. Meixun Zhao and Jeffrey Bada of the Scripps Institute of Oceanography in San Diego suggested that a definitive test of which model (volcanoes or impact) was correct would depend on whether evidence could be found for the presence of amino acids in the K/T boundary layer. Amino acids are key organic molecules used to make proteins and are found in all living things. However, certain types of amino acids are only found in meteorites.

Amino acids come in two forms, which are essentially mirror images of one another; for example, alanine can occur as L-alanine or D-alanine. The point is that terrestrial life uses essentially only the L types, yet the D variety has been found in abundance in meteorite samples, suggesting an extraterrestrial and non-biological origin. If these amino acids were present in the K/T sediments, argued Zhao and Bada, it would prove unambiguously that a cosmic collision, either with an asteroid or with a comet, had been involved in bringing them here. No one was going to argue that the heat and violence of a volcanic eruption would introduce peculiar amino acids into the atmosphere. The heat of a volcano would destroy such molecules.

When they studied the K/T deposits at Stevns Klint in Denmark in 1989, Zhao and Bada found the evidence they were looking for. Both types of amino acid were identified in deposits associated with the K/T boundary, but not directly with the iridium peak in the K/T fireball layer. The alien amino acids were found just above and below this layer.

In a related study reported in 1990, Kevin Zahnle and David Grinspoon of the NASA Ames Research Center in California suggested an ingenious explanation for this apparently anomalous situation. The amino acids were deposited with the dust from a comet trapped in the inner solar system. They wrote that "Amino acids or their precursors in the comet dust would have been swept up by the earth both before and after the impact, but any conveyed by the impactor would have been destroyed."

The impact event would have created far too much energy to allow amino acids to survive, but for 50,000 years before and after the impact, amino acids linked to cometary dust rained onto the earth. The relative fraction of amino acids around the iridium layer in the K/T boundary deposits is much higher than in the Murchison meteorite, which fell to earth in Australia in 1960, one of the objects known to contain alien amino acids.

This scenario leads one to think about what might happen when a comet merely passes close to the earth without smashing into the planet. Close encounters happen more often than direct hits and comets passing in the night could introduce the stuff of life from space into the terrestrial environment. Such space dustings would barely be recorded in geologic layers, and would be very difficult to identify without an extraordinary amount of painstaking work to interpret the chemical makeup of otherwise ordinary looking deposits. Yet the effect on life of repeated injections of comet dust containing alien biological molecules into the atmosphere may have been important in evolutionary history.

This idea is reminiscent of a suggestion made in a number of books by Sir Fred Hoyle and his colleague Chandra Wickramasinghe in Great Britain, who claim that there is life in space and that comets play a role in "infecting" earth with key molecular, and possibly even bacterial or viral, material. The typical response to their arguments has been to point out that it is a big step from complex organic molecules known to exist in comets to viruses. The discovery of alien amino acid deposits around the K/T boundary may do something to bridge that gap.

To return to the story revealed by the K/T boundary clay, it contains even more evidence that a violent impact must have been involved. In 1991 Haraldur Sigurdsson of the Graduate School of Oceanography at the University of Rhode Island together with several colleagues published a report on the nature of tektitelike glass preserved in what geologists call the Belloc layer, in Haiti (Figure 1-5). The physical structure and chemical composition of this type of glass requires a violent impact and completely rules out a volcanic origin.

Tektites are glassy objects, often the size of small pebbles, that are found spread over various regions around the world. They were formed when molten rock solidified while traveling rapidly above the atmosphere. Several distinctly different forms of tektites are found in a number of "strewn fields." One of them covers Australia and a part of the bottom of the Indian Ocean. That field is littered with tektites (Australites) deposited about 700,000 years ago. Another strewn field covers parts of North America, including Texas, which is about 33 million years old. The most widely accepted explanation for the origin of "normal" tektites is that they solidified while traveling above the atmosphere after having been molten and splashed into space by the violent impact of a comet or asteroid.

The tektites in the K/T boundary layer are in the form of tiny glass spheres only a tenth of an inch in diameter. They are silica-rich with a chemical makeup expected from the melting of continental crustal rocks. The glass also contains a great deal of calcium. These microtektites were apparently formed by an impact on a continental crust overlain by marl sediments. (Marl consists of loose aggre-

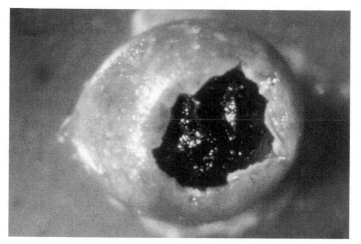

Figure 1-5 A partially altered tektite found in the K/T boundary layer near Beloc in Haiti. The thin shell surrounding the interior dark, glassy tektite is weathered tektite material that has turned to a claylike substance called smectite. Such objects were discovered in the lower few centimeters of half-meter thick bed of impact ejecta. The discovery of tektites in the same bed with an iridium anomaly and shocked quartz grains is powerful confirmation of the idea that a large asteroid or comet struck earth 65 million years ago. (Courtesy Glen A. Izett, U.S. Geological Survey, Emeritus)

gates of earth and clay that contain a lot of calcium carbonate or limestone.) This tektite composition, combined with a lack of crystal structure, rules out a volcanic origin.

In their report, Sigurdsson and colleagues suggested that the Manson structure in Iowa, a craterlike geological uplift, was a candidate for the impact site. However, unbeknownst to them, evidence was gathering that a hidden depression in the Yucatan in Mexico was the crater everyone was looking for. As will be related in the next chapter, by 1992 the picture was complete. A crater beneath the town of Chicxulub in the Yucatan marks the place where at least one K/T comet slammed into the earth 65 million years ago. Back then the Yucatan Peninsula lay beneath 100 meters of water and the impact produced landslides that triggered tsunamis (enormous waves) that together with the direct consequence of the crash literally emptied the Caribbean, for a while at least. When the water sloshed back in it must have created a further stunning upheaval around its shores.

In 1993 E. Robin and colleagues in France reported that their analysis of deposits in the Pacific Ocean suggest that another object, perhaps 2 kilometers across, must have slammed into the earth at about the same time as the Chicxu-lub event. As we shall see later, it appears increasingly likely that the K/T killer did not, in fact, act alone. The object that punched a hole in the Yucatan 16 kilo-meters deep may have been the largest of a group of objects that slammed into the earth back then.

In the face of the ascendancy of the impact theory for the K/T extinction event, the volcanic theory has all but gone extinct. To paraphrase Baron Cuvier, 65 million years ago life on earth was interrupted by a particularly frightful event when it was struck by a large object from space. The K/T mass extinction event was precipitated by the consequences of that impact. At least one of the craters has been found, and computer simulations of what happened then paint a vivid picture of catastrophe, one that can best be appreciated by adjusting the way in which we look at the earth and its environment, and at the evolution of life over long periods of time.

2

THE SAGA OF THE CHICXULUB CRATER

WHEN the Alvarez team announced to the world that the K/T boundary clay contained a excess of iridium they suggested that it could only be explained if a comet or asteroid had slammed into the earth 65 million years ago. The iridium was deposited when a cloud of debris created by the vaporization of the object upon impact girdled the earth and fell back to form the so-called fireball layer. Most earth scientists were skeptical when they first heard about this. If an object 10 kilometers across had collided with enough force to trigger a global environmental catastrophe that precipitated the extinction of more than half of the species alive at the time, where was the crater? It didn't take crater experts long to figure that the scar left by such an impact should be huge hole in the ground about 180 kilometers across and a tenth as deep. If it existed, it shouldn't be hard to find, unless it was under the ocean somewhere, or covered in vast amounts of sediment. It turns out that when the search for the crater began there were several people, perhaps dozens, who already knew where it was. However, they either didn't know that the search was on, or weren't allowed to reveal what they knew.

The saga of the discovery of the K/T impact crater beneath the north coast of the Yucatan Peninsula of Mexico began many decades before the discovery of iridium in the K/T boundary layer. The saga goes all the way back to 1947 when a gravity survey was started in the Yucatan by the Mexican national oil company, PEMEX. Surface gravity measurements allow geophysicists to detect the structure of rock formations deep beneath the earth's surface. The study of gravity maps of a region then helps the scientists to figure out where oil might be found; at least that is the goal. The Yucatan survey turned up some intriguing data, including hints of a circular feature some 1,000 meters deep. In the early 1950s test wells were drilled, but no oil was found.

One of the first wells was drilled at a town called Chicxulub (pronounced Chic-sa-lube) between Progreso, at the coast, and Mérida, the inland capital of the region. When the well reached a depth of about one kilometer geologists noticed that there was something odd about the core samples being brought up. No one really knew what they had run into, and no one had time to stop to think about it. Oil exploration is a highly profit-oriented business and detours to think about the scientific implications of what was found were not encouraged. Even if the geologists had stopped to think about what they'd found, it is highly unlikely that any of them could have imagined what they had stumbled into; a 65-million-year-old impact crater, the scar left by the object that precipitated the extinction of the dinosaurs. In 1952 when the existence of the strange gravity anomaly was first confronted, the notion of impact craters on earth was all but unheard of. In the 1960s and 1970s, however, the American space program made scientists aware of the widespread existence of impact craters in the solar system (see chapter 12).

The core samples from Chicxulub included fragments of a type of rock called andesite, usually associated with volcanic activity. The fact that there were no volcanoes anywhere in the region did not prevent the volcanic interpretation of the nature of this material to gather a life of its own. Three decades later experts still argued that the cores had come from a buried, extinct volcano.

Because PEMEX found no oil in the Yucatan, the wellheads were sealed, and the geophysicists moved on to explore other areas of the country. Over the next decade the nature of the gravity data continued to fascinate PEMEX workers and contractors. In 1966, Robert Baltosser was working for a company called Seismographic Service Corp. in Tulsa, a contractor employed by oil companies to do gravity survey work. He was approached by a PEMEX manager and asked whether he might help them reconsider, and perhaps reanalyze, the Yucatan gravity data in order to figure out what they had found. The PEMEX employees were still puzzled and thought that a fresh mind might cast new light on the nature of the peculiar structure in the gravity map. Baltosser recalls that when he was shown the data he realized he could not give a quick answer and would need to look closer before even being able to decide whether further study on his part

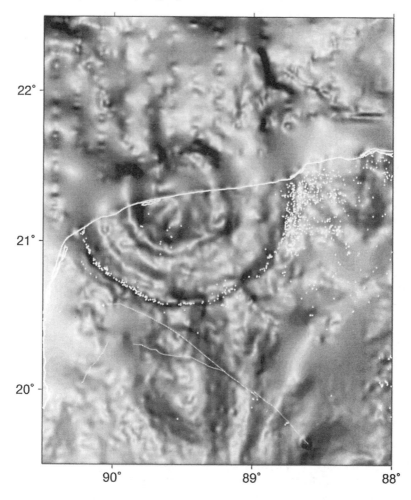

Figure 2-1 The Chicxulub crater as revealed on gravitational maps. The image shows the gravity structure for the northwest corner of the Yucatan Peninsula. The buried Chicxulub crater is revealed as a striking series of nearly concentric rings. The locations of the coastline, cenotes (water-filled sinkholes), and recent faults are shown in white. Areas lacking data appear blurred, and offshore ship tracks make broad, linear features in the plot at upper left. The numbers along the edge of the map are longitude (horizontal, at bottom) and latitude, along the left side. (Courtesy A. R. Hildebrand and Geological Survey of Canada)

was merited. The PEMEX manager suggested he take the maps with him, think about how to proceed, and let him know.

Within days of arriving back in Tulsa, Baltosser called his PEMEX contact. "I don't think I can offer you much help," he said. "There was a major meteor strike in the area." His analysis was that the gravity data were severely compromised by the presence of an enormous crater (see Figure 2-1). At the time, Baltosser had just done work on the gravity signature of the Wells Creek structure in Tennessee, which had been identified as an ancient impact crater. To him it seemed that the gravity signature in the Yucatan was very similar, except that crater was much larger, more like 100 kilometers across compared with 13 kilometers at Wells Creek.

A few days later he got a call from a colleague in the Tulsa company who told him that his contact at PEMEX had been fired as a result of an organizational shake-up. Furthermore, a stricter corporate policy about sharing proprietary data had been instituted and Baltosser was advised to immediately return the gravity map, which he did.

Because the data obtained during oil exploration have potentially great economic value, they are essentially kept within the company doing the work. The data are said to be proprietary, which means "classified." Since PEMEX had the exclusive right to look for oil in all of Mexico, it was unlikely that any outsiders, except contractors hired by the company, all of whom had to sign a confidentiality agreement, would ever get to see the data Baltosser had stared at in amazement when he realized a huge crater lay deep beneath the Yucatan peninsula. He could speak no further about his hypothesis.

Meanwhile, in the early 1970s, drilling resumed in the Yucatan. Since then, the adjoining Campeche Platform has become the center of Mexico's oil wealth. While this work was proceeding, PEMEX employees and consultants continued to talk among themselves about the ringlike gravity anomalies. Old-timers recalled finding odd looking rocks back in the 1950s and the andesite that had naturally, but erroneously, been blamed on volcanism, and the myth of an old volcano beneath the sediments of the Yucatan continued to grow. History would show that the cores from the 1970 drillings were ultimately stored in a warehouse in Coatzacoalcos in the state of Vera Cruz.

In 1978 a key player entered the scene. Glen Penfield was working for a U.S. contractor and was given the task of quality control in a large-scale aeromagnetic survey over the waters of the Gulf of Mexico, just north of the Yucatan over Mexico's territorial waters. Measurements of the strength of the earth's magnetic field with sensitive magnetometers reveal small changes produced by rock formations that are magnetized in slightly different ways, or which vary in depth over a terrain. Just as is true for gravity studies, geophysicists use such data to make maps of the location of magnetized rock structures deep beneath the ground. This gives them further clues that would hopefully aid in the search for oil.

Penfield was working under the same restrictions as any PEMEX employee or contractor. Whatever he found had to remain proprietary. In his case, what he found was stunning.

The magnetic survey was carried out from a plane flying back and forth 5,000 meters above the sea along of 400 kilometer traverses in the east-west direction. Parallel traverses were separated by about 4 kilometers. On some days they would fly north-south traverses separated by 20 kilometers This allowed the measurements, which were recorded every 30 meters, to be tied together to make a map of the magnetic field in the region. (Geophysicists measure the earth's magnetic field strength in units called nanoteslas, after the famous eccentric inventor and experimenter with electricity, Nikola Tesla. The average earth's field is 35,000 nanotesla and the instruments used in the magnetic survey could detect tiny variations as small as half a nanotesla.)

The data were collected and displayed on a paper chart recorder (known as an analog record), and digitally on magnetic tape for later analysis. Penfield was supposed to check the quality of the data by looking over the analog output, which came in the form of about 15 meters of paper per 400 kilometer traverse. It was while looking at these data that he first noticed a tiny glitch north of Progreso, over the sea somewhere. This indicated a 2 nanotesla signal of a type that would usually be ignored. Such signals could result from sudden changes in the earth's magnetic field produced when a wind gust of charged particles that blows out from the sun slams into the magnetic field of earth.

What puzzled Penfield was that when the plane reversed its flight path and flew a new traverse, another glitch occurred more or less over the same part of the sea. Curious, he began to make a sketch of where the glitches occurred in successive flights and plotted them on a map of the area.

Over the course of the next few weeks his map of the glitch location showed a semicircular structure. Then, in April, after a month of survey work, the plane flew over a region near the coast where an enormous signal of 600 nanotesla suddenly appeared. In subsequent traverses it reached 1,000 nanotesla, 3 percent of the earth's normal magnetic field.

When he plotted all the data, Penfield realized he was staring at the signature of part of a hidden crater. The large peak was reminiscent of the typical central uplift in large impact craters. It had a very strong magnetic field and it was surrounded by an outer rim where the magnetic field was much weaker, yet well above the background value for the region.

When the survey flights paused briefly in order to move the transponders that were used to pinpoint the location of the plane during its traverses, Penfield got talking to colleagues at the site, who told him about the gravity data that had been obtained decades before. Someone dug out a copy of the map and, when he put it together with his hand-drawn magnetic field map, he saw that the two sketched out a perfect circle with a peak at the center. Now there was no doubt in his mind.

"It was a huge, perfect, fossil impact signature," he recalls, reliving his moment of discovery. "The fine structure, the semiconcentric lobe features, they were clearly visible."

"This was one of the greatest moments of my life," he added. He experienced the wonderful thrill of discovery that drives so many scientists, the marvelous feeling that comes from seeing or discovering something that no one has ever seen before. He was staring at an impact crater lurking beneath 1,000 meters of sediment. This had to be important, he thought, but he couldn't talk to outsiders about it.

He had no doubt that the crater in the Yucatan was of impact origin. He had already flown over many volcanic craters and the magnetic signature of this feature was nothing like a volcano. Volcanoes simply did not have the outer ring that is a mark of an impact structure. As he says, "This was a textbook example of an impact crater and it was unspeakably exquisite." The problem was that no textbook had examples as good as the data at his fingertips.

Several years later when the data tapes were sent to PEMEX headquarters they were processed to highlight the existence of large-scale structures beneath the ground by "normalizing" the data to a standard altitude of 4,000 meters as opposed to the 500 meters at which the plane had flown. This process is called "upward continuation" by geophysicists and it allows them to more readily determine the nature of large-scale changes in the magnetic field signals, changes that would indicate the presence of large domes of material far beneath the surface, for example. The procedure removed all traces of rapid changes in magnetic field strength limited to small areas, such as the glitches that had fascinated Penfield.

After this normalization procedure was finished, the original data tapes were inexplicably lost. Penfield's claims that the signature of an impact crater existed in the magnetic field maps now rested on the still proprietary analog records and his sketch maps. In August 1978 he wrote a report for PEMEX in which he suggested that the structure was an impact crater. The report was filed.

In 1980 Penfield came across the announcement by the Alvarez team of the discovery of iridium in the K/T boundary clay and suspected immediately that the Yucatan crater had to be the "smoking gun." Its initial dating had placed the sediments above the crater at about 80 million years, but if it was a crater, those deposits could have been churned up a lot. The crater could be a little younger, depending on the way the impact ejecta were blown about. To him the 65-million-year date of the K/T boundary clay was close enough to erase any doubt in his mind. He recalls that he immediately wrote to Walter Alvarez to suggest that the crater lay in the Yucatan and never heard back.

In 1981, aware that the existence of the Yucatan crater had to be announced to a larger audience, he convinced his Mexican colleague and project supervisor, Antonio Camargo, that they should present a report on the crater at a meeting of Society of Exploration Geophysicists in Los Angeles. Camargo agreed and in early October they informed 30 to 35 colleagues of the likely existence of a very large impact crater beneath the Yucatan. In that same week a group of planetary scientists, paleontologists, and geologists were meeting in Utah at the first

"Snowbird" conference on the nature of mass extinctions and impacts. Key in their minds was the task of finding the crater. Neither group knew of the other's existence. (During my research into this story, I learned from Robert Baltosser that PEMEX officials were somewhat upset that their proprietary data had been discussed in Los Angeles.)

Now the saga takes an unusual twist. A journalist at the Houston *Chronicle*, Carlos Byars, picked up on the crater story and on Sunday, December 13, Penfield and Camargo's claim hit the front page. "Mexican site may be link to dinosaur's disappearance" was the headline. A map of the Yucatan coastline showed the location of the crater centered beneath the town of Progreso, a map that is essentially an outline of what is shown in Figure 2-1.

The news got no further, however, not for a while at least. How Byars came to write the story is a fascinating coincidence and he was later to facilitate a crucial meeting of two key players that lead to the eventual, widespread recognition of the relevance of this crater.

The reason that Byars did the story was that he had been working in the oil industry as a writer. He knew Glen Penfield, who was associated with a sister company. One day Byars had visited Penfield's office where he was shown the gravitational and magnetic data. "Glen asked me what I thought I was looking at," he recalls. "I said it looked like an impact crater, but I wasn't telling him anything new."

He appreciated that Penfield could not publish his finding but a journalist might be free to write a story after Penfield and Camargo had spoken at the Los Angeles conference. By the time that happened, Byars had taken a job at the Houston *Chronicle*.

At about the time Penfield was trying to draw attention to the existence of the Yucatan crater, Robert Baltosser read an article in *Astronomy* magazine about the rare metal iridium being found in the K/T boundary layer around the world, and that the search for the crater was on. He, too, and quite independent of Penfield (neither knew of the other's existence), suspected that the Yucatan structure was the likely site for the ancient impact. In January 1982 he wrote to the author, Richard Teske at the University of Michigan, to tell him about this possibility and added that if anyone was interested, he, Baltosser, would be willing to work to get the gravity data released, if that would help prove the point. Teske passed along the information to a local geologist who concluded that the impact hypothesis, while intriguing, seemed unlikely to be correct. After further work on his own part to learn more about the Yucatan gravity data, which included writing to Mexican government officials for information and searching for satellite images of the area, efforts that led nowhere, Teske put the issue out of his mind.

Later that year the story of the crater was picked up by *Sky and Telescope* magazine, which, in March 1982, included a brief report in its news section. This announcement of the existence of an impact crater in the Yucatan, the likely remnant of the K/T collision, did not seem to stir the planetary science community engaged in the search. (The editor in charge of that column now wonders whether anyone read it at all.)

Meanwhile, Penfield was growing more obsessed with his crater and he went to talk to crater experts recently spawned by the U.S. space program. Several of them worked at the Johnson Spaceflight Center near Houston and his contact there, Bill Phinney, stressed that Penfield's hypothesis could only be taken seriously if cores from wells were located and analyzed for evidence of shocked material associated with the violence of impact.

That sent Penfield on a personal odyssey for samples. If he could track down any of the cores brought up from the Yucatan wells, he would be on his way to providing the Johnson scientists with material they needed to test. He soon discovered that the cores from the 1970s drilling had been stored in Coatzacoalcos, but that the warehouse had been burned to the ground. Undaunted, he traveled to the site and found that the ruins had been bulldozed flat. This visit was motivated by a small hope. He knew that the entire surface of the northern Yucatan Peninsula is composed of whitish carbonate rocks; if he was to find dark andesitic rocks in the ruins of the warehouse, they would surely have had to come from the crater area. As Penfield told me, there are no andesitic rocks found within 600 kilometers of Mérida and there is no surface water or even rivers of any kind in the region to transport such rocks from some distant, volcanic region to the Yucatan.

At this point we must make a small digression as regards the relevance of the andesite. It is known as a melt rock, produced when ordinary sedimentary rock is melted, usually as the result of volcanic activity. Therefore, it was quite reasonable that the earliest interpretation of the presence of andesite in the Yucatan cores suggested ancient volcanism. But andesite is also produced when the center of an impact crater is melted by the violence of the collision, and the only certain way to tell the two forms apart was to find evidence for shocked quartz. That would indicate an impact origin and would rule out volcanism.

The next phase of Penfield's quest was undertaken in his spare time. He would search for the dark rocks at the wellheads, just in case someone had left clues lying around after they cleaned up. He flew to Mérida, hired a taxi, and told the driver to first take him to Sacapuc, the town where one of the wells had been drilled. He got out with his shovel and pick and stopped strangers to ask to be directed to the oldest person in town. He was led to an old woman who said, yes, PEMEX had drilled there in the 1950s, and that the area around the wellhead was now fit for only one thing—a pigsty. Still undaunted, Penfield found the site and began to dig around in a foot of pig manure in the heat of a Yucatan day to search for fragments of old cores that might still be lying around. His only success was a keener appreciation of the nature of pig manure.

His next stop was Chicxulub. "A beautiful, romantic location," he says,

Figure 2-2 Glen Penfield and his Maya guide upon chopping through the under-growth to reach the wellhead at Chicxulub where he sought, in vain, for samples of rock from the drill cores that might be analyzed for evidence of shocked quartz that would prove the crater, over 1,000 meters below this spot, to have been pro-duced by impact. (Courtesy Glen Penfield)

recalling the sight of green tree snakes decorating the area to which an old man, already drunk at 10 o'clock in the morning, led him, a place mercifully free of pigs. Together they chopped through undergrowth with a machete to reveal the wellhead he was looking for (Figure 2-2). There he noticed debris spread about which made him wonder whether someone else had been rooting about in search of clues as to what lay a kilometer below. He found some fragments of material from old cores that appeared to have been mixed with cement to fill the well. He carried 20 pounds of this debris back to Houston and handed it over to the peo-ple at Johnson Spaceflight Center for them to analyze for clues as to the nature of the underlying crater. Unfortunately, by then the Johnson scientists were apparently convinced that the structure was a volcano and didn't take the quest any further. As Penfield speculates today, "The myth of volcanic origin was so well established that nothing could shake it." As regards his own involvement, duties took him elsewhere and it would be eight years before he was drawn back into the saga.

Even as this was going on in the early 1980s, Carlos Byars, the Houston *Chronicle* journalist, began attending the lunar and planetary science conferences held at the Johnson Spaceflight Center on an annual basis. He had been doing stories about the K/T impact event and extinction of species and had heard many presentations about potential candidate craters that were supposed to mark where the asteroid or comet had slammed into the earth 65 million years ago. At every conference he buttonholed some of the scientists and pointed out that there was a crater not on their list that merited a closer look. Every year he was apparently ignored. "They regarded me as a good, competent science writer, but no expert on craters," admits Byars.

In 1988 the Second Snowbird conference was held in Utah and still none of the key people took much notice of Penfield and Camargo's crater, even if they had heard about it, usually from Carlos Byars. It appears that the volcano hypothesis exerted a strong influence over the planetary scientists. I also suspect that the professional scientists were reluctant to pay much attention to "outsiders" such as Baltosser, Penfield, and Byars, a syndrome that is common within science, understandably so because a great deal of the "information" brought by amateurs to a specialist in a given field tends to be irrelevant. An exception has occurred in astronomy, where amateurs have made great contributions, but even there they often have to fight to be heard.

Byars recalls that at one of the lunar and planetary science conferences a scientist did listen to him and promptly handed the suggestion to a graduate student for further investigation. The student dismissed the notion. "They've probably been kicking themselves ever since," Byars now says.

In early 1990 Alan Hildebrand, then a graduate student at the University of Arizona, met Byars and heard his story. Hildebrand was on his own quest of the K/T crater. He had been doing field work in Haiti where together with colleagues he studied a thick layer of ejecta, in places more than a meter thick and known as the Beloc layer, by then known to be clearly associated with the K/T impact event because, in part, of the rich harvest of tektites found there. Ever since Jan Smit of the Geological Institute of the Free University in the Netherlands had pointed out that around the Caribbean the fireball layer was deposited on top of this ejecta layer, it was suspected that the crater would likely lie in that region of the world. The ejecta layer refers to the ejected material that fell back to earth relatively close to the impact site, whereas the fireball layer, referred to in Chapter 1, contains the excess iridium that blankets the earth.

Hildebrand knew about less thick ejecta blankets in Texas and was able to conclude, on the basis of how the thickest layers were found around and in the Caribbean, that the crater had to be somewhere in that area. He was working on a report about this study with his colleague William Boynton in which they suggested that the crater might lie in what is called the Colombian basin beneath the Caribbean, 2,000 kilometers south of the Yucatan Peninsula. A watery impact site had been inferred from the nature of the ejected material whose chemical composition contained clues that strongly suggested to them that the impact had occurred on oceanic crust—in other words, in the sea or ocean.

A further clue to a watery impact came from evidence in the geological record that great waves had once washed over the southern part of North America and what are now the Caribbean islands, waves that had been generated by the violence of a cosmic impact, enormous tsunamis (see chapter 12).

The report written by Hildebrand and Boynton about the possible location of the impact site in the Colombian basin had already been sent to a journal (*Science*) when Carlos Byars found Hildebrand's receptive ear. As Hildebrand now recalls, he may have been one of the last active K/T researchers actually to learn about this crater. History will, however, recount that he was the first academic research scientist to actively pursue this lead.

Hildebrand quickly called Penfield, who recalls that "I told him that I had this impact crater in the Yucatan that was as plain as the nose on my face." Hildebrand recalls it differently. "When I first talked to him he told me that they [Penfield and Camargo] weren't sure it was a crater, and that they hadn't done anything with it for years."

Hildebrand and Boynton did include a reference to the Penfield and Camargo claim in their *Science* paper at the last moment before going to press: They had considered several possible sites in the Caribbean and added, "Alternatively, the [approximately] 200 kilometer diameter, circular, magnetic and gravitational anomalies reported by Penfield and Camargo on the continental Yucatan Platform could indicate a buried impact structure...."

Soon after his discussion with Penfield, when they appreciated that to make their case they would have to find "unclassified" data to prove the existence of the crater, Hildebrand visited Ottawa for a job interview at the Geological Survey of Canada. There he was able to look at gravity maps of the Caribbean and saw that there was no likely crater in the Colombian basin and that the Yucatan structure was visible, albeit in less detail than the gravity and magnetic field data that remained proprietary to PEMEX.

In April 1990 they sent a report to *Nature* proposing that the K/T crater lay hidden beneath more than 1,000 meters of sediment in the Yucatan and an anonymous referee promptly rejected the paper. This is one of the hazards scientists face in order to publish their results. A paper gets written and sent to a scientific journal, which then sends it out to an anonymous referee who expresses what will hopefully be an informed opinion about whether the paper is up to the standards of the journal and should be published. Authors are then usually encouraged to make changes suggested by the referee, after which the paper will be reconsidered. In the case of the Hildebrand and Penfield's contribution, it was rejected outright. It is one of the perils of the business that sometimes critical papers get rejected for what later turn out to have been sincere, although occasionally, misguided reasons. (Sometimes papers are rejected for very much more personal reasons, and tales of such instances abound within the scientific community.)

Hildebrand, like Penfield before him, knew they would not be able to prove their point without finding andesite samples of well cores from the Yucatan to see if they contained shocked quartz. So he began to ask around and learned from two or three people that cores from the early 1970s had been sent to Al Weidie at the University of New Orleans who was interested in underground water in the Yucatan. They had been sent by a PEMEX official who was a close personal friend.

Penfield happened to be traveling to New Orleans so he paid a visit to Weidie and was able to sort through 600 boxes of samples. He recalls seeing one package dated 1963, which suggested to him that at least some of the samples dated from the wells of the 1950s.

Penfield shipped off some of the New Orleans cores to Hildebrand, whose colleague, David Kring at the University of Arizona, almost immediately demonstrated evidence for shocked quartz in the samples, a sure sign that the crater had been produced by a violent impact. An ever growing team of collaborators then wrote a report about this and tried to get it published in *Nature*. It was again rejected by, as it later turned out, the same anonymous referee whom the intrepid crater hunters succeeded in identifying as well. The paper was finally published in the journal, *Geology*, in September 1991. The authors were Hildebrand, Penfield, D. A. Kring, M. Pilkington, Camargo, S. Jacobsen, and W. V. Boynton.

Gradually more crater core samples turned up from various places. It turns out that many had been hoarded for decades by curious geologists who had wondered about their nature. These "souvenirs" were suddenly of crucial importance to the understanding of what struck the Yucatan.

By 1991 the identification of the crater as the K/T impact scar was complete. Carl C. Swisher from the Institute of Human Origins in Berkeley and his colleagues used radioactive dating (K_{40} to Ar_{40}) to establish an age of 65 million years for both the deposits inside the crater and the tektites in Haiti that at one point had drawn Hildebrand to that area.

Glen Penfield looks back on the turmoil and thinks that "the discovery of the crater has been a triumph of geophysics." It was found not because someone fell over the crater at the surface, but because gravity and magnetic field data showed its presence. He hopes that someday someone with a half million dollars to spare will fund a careful, low altitude, magnetic survey of the region. "This will reveal exquisite detail of what an impact crater looks like," he says. The asteroid or comet slammed into a few miles of unmagnetized sedimentary material that covered an unweathered layer of magnetic rock (igneous). As a result of the shattering impact explosion, the ground peeled away and material that fell back, the ejecta, immediately covered the entire area. At the same time, a central peak rebounded from a depth almost to the earth's mantle. The crater has lurked down there ever since, utterly undisturbed for 65 million years. Penfield thinks that "this will become one of the great natural [geophysical] laboratories of the 21st century."

Figure 2-3 A typical cenote in the Yucatan, a sinkhole now filled with water, which is used by locals for recreation. Many of the cenotes are arrayed around the edge of the crater that lies far beneath the surface, as shown in Figure 2-1. (Courtesy Glen Penfield)

In 1991 Kevin Pope of Geo Eco Arc Research in La Canada, California, and two colleagues, Adriana Ocampo and Charles Duller, added another twist to the story. The Yucatan is pockmarked by sinkholes called cenotes; a typical cenote is 50 to 500 meters in diameter and contains water with depths of 2 to 120 meters. Figure 2-3 shows a typical cenote. It turns out that their location defines the outer rim of the crater, as is shown in Figure 2-1. The sinkholes were apparently formed by collapse of material around the crater rim. Because surface water is nonexistent in the region, the sinkholes were of fundamental importance to the Mayan civilization. Without the K/T collision, the territory would almost certainly have been unfit for habitation.

When I visited this area while on vacation in 1980 and stood beside a large cenote at Chichén Itza, which lies just outside the main rim of the deeply buried impact crater, a fact of which I was quite oblivious at the time, my imagination was stirred by a guide's tale that human sacrifices, mostly to the Mayan rain gods, were common at the more sacred of the sink holes. Whether or not the human sacrifices were virgin beauties, as the guide claimed, I do not know, but it is fascinating to speculate that such rituals were held at locations determined by geological processes mediated by the K/T impact.

Today the K/T crater is known by the name of Chicxulub. I asked both Penfield and Hildebrand how the name was chosen. They recall it differently. Penfield: "We picked the name together over the phone. I liked the idea of the

Chicxulub which is almost unpronounceable to English-speaking scientists. Also I like the notion of its meaning, 'the tail of the devil' or 'the backside of the devil.' He remembers that Hildebrand called him with a choice of three names; Mérida, Progreso, and Yucatan and that, up to that time, they had sometimes referred to it as the Progreso crater, since Progreso is directly above what was once ground zero. But the relatively unpronounceable Chicxulub appealed to Penfield, partly for "impish" reasons.

Hildebrand remembers it differently: "I called Penfield for advice and had three names in mind: Mérida, Progreso, and Chicxulub. Glen said any of the three would be fine. Then I called Richard Grieve to find if there was some rule for crater naming. He said no and gave no opinion as to what name should be used."

Does Chicxulub actually mean "tail of the devil"? Hildebrand says that he has been unable to find anyone who can confirm this claim. He has asked Maya experts who do not recognize the word at all. "Sign of the horns" was inferred by one, but that may not be correct either. He asked local Mayans and was amused by the best suggestion he heard: "The ticks are really bad here."

So who discovered the crater? There is no simple answer. It seems as if Robert Baltosser was the first to categorically state that the structure was an impact crater. However, no written documentation of this claim exists prior to the 1982 letter he wrote to the author of the article in *Astronomy*. But by then Penfield and Camargo had already gone public with the idea. They spoke about it in October 1981 and the Houston *Chronicle* article makes it clear that they were suggesting that this was the crater associated with the K/T event. There is also no doubt from a *Sky and Telescope* report published in March 1982 (p.249) that this was the case. But in the academic world, claims of priority are not so casually accepted. The academics wanted proof in the form of the shocked quartz. That proof was formally obtained in the laboratory by David Kring working with the core samples laboriously unearthed by Hildebrand and Penfield and the results published by a collaboration of authors—yet they had not really *discovered* the crater.

Penfield and Hildebrand were both instrumental in broadcasting the fact of the existence of the crater to a larger audience. Penfield deserves credit for having braved ten years of scoffing by professional scientists, as well as deposits of pig manure, in his quest, and Hildebrand for rallying colleagues and completing the quest for core samples that allowed the K/T connection of the crater to be established.

Establishing an identity or finding an explanation is not the same as discovery. A dramatic example in astronomy I think parallels the saga of the Chicxulub crater. When Robert Wilson and Arno Penzias discovered strange radio waves from all over the sky, they published a paper that did little more than announce the existence of this so-called cosmic microwave background. They did not attempt to explain it. Nor were they the first to observe the signal. But they were the first to sample the distribution of the signal around the sky and found it to be uniform. For that they received the Nobel Prize in Physics in 1965. The astronomers who immediately explained the origin of the signal (the Big Bang that set the universe into existence) and who published the explanation in the same issue of the journal where Penzias and Wilson announced their result did not share the prize.

Penfield and Camargo were the first to publicly announce the presence of an impact crater beneath the Yucatan. Before them, Robert Baltosser had said the same thing, but had been restrained by confidentiality from publishing anything. After them, the identity of the crater and its impact nature entered the scientific mainstream and arguments still rage as to who deserves the credit for discovery. The full story remains to be told, a story of personality conflicts and political infighting that go way beyond anything I can delve into in this book. As Hildebrand says, "The full story is a loaded subject" and my attempts to communicate what appears to have happened will not satisfy all those who were involved. Such is the nature of historical research. Reaching for the facts even only a few decades after events in which the participants were not aware of the fact that they were making history is notoriously fraught with difficulty. If they had been aware of the importance of their work, they might have kept journals, always a good idea, I think.

This story has more parallels in other areas of scientific discovery. Key discoveries are not always attributed to the first person who observes a phenomenon, but to the first one who asks the right question about its nature and who has the good fortune to find the answer. I have written about this in the context of the discovery of radio waves by Heinrich Hertz, at least the sixth person to observe the phenomenon. (See *Hidden Attraction*, referred to in the bibliography.)

The Chicxulub story has many of the elements of lost opportunities and luck that are the hallmark of other major scientific discoveries. Robert Baltosser, whose original suggestion about the likely impact feature in the Yucatan first went unheeded and unpublished, now estimates that by the time 1990 rolled along at least 50, perhaps 100, geologists and geophysicists knew about its existence.

As Alan Hildebrand wrote to me: "How much simpler K/T research would have been if Robert Baltosser or Glen Penfield had been able to publish some of the confidential data at their respective times. When Alvarez [and his colleagues] did their work in 1979 they would have had a candidate [crater] immediately." Instead, the planetary scientists waited 12 years before the connection was made, a connection that came 24 years after Robert Baltosser's first insight.

So what did hit the Yucatan 65 million years ago? Was it an asteroid or a comet? For that matter, what are asteroids and comets?

3

Solar System Debris

*T*HE object that slammed into the earth to precipitate the dinosaur demise was no stranger to the solar system; it had been lurking about its outer regions ever since the sun and planets formed 4.5 billion years ago.

Like a construction site littered with builder's materials after the work is done, debris left over from the formation of the sun and planets is scattered throughout the solar system in the form of comets and asteroids. From among this population the great impactor that triggered the K/T extinction event originated. Unfortunately there is no way to cart the debris away so that earth won't smash headlong into another comet or asteroid. Whether we like it or not, we live with the hazard of occasionally finding some of this stuff directly in the path of the earth's orbit around the sun. In fact, every day something from space slams into our planet.

The generic name for icy, dusty, rocky, or metallic objects that wander about in interplanetary space is "meteoroid." After the process of planetary formation was essentially complete, a great deal of meteoroidal material was left over. Depending on its fate, a meteoroid hurtling into the earth's atmosphere today earns a new name. It is called a *meteor* in the case of a tiny pea-sized particle that burns up in the atmosphere to produce a momentary fiery trail known as a shooting star. These have long been the focus of superstitions because of their obvious associations with the heavens and, therefore, with gods that might reside there. Even in our time, it is common to "make a wish upon a star" when a meteor is seen. Every night you can see dozens if not hundreds of meteor trails. On a good night in a clear location half a dozen an hour is about as good as you can expect.

In addition to the meteors that burn up to leave a visible trail, there are countless that are too small to be seen, enough to allow tens of millions of wishes a day. These heat the atmosphere enough to produce trails of hot gas that reflect radar signals.

Most of the meteor-sized particles are left over from the gradual breakup of comets and asteroids. At specific times of the year a hail of shooting stars may decorate the night sky. These are called *meteor showers* and a good one provides a spectacular sight (Figure 3-1). They occur annually when the earth runs into the orbits of comets that broke up a long time ago and scattered debris along their paths. Table 3-1 lists the well-known meteor showers, when they are likely to be seen, and the names of their parent comets.

All in all, about 10,000 tons of space debris falls to earth every year, mostly in meteoric form. But sometimes a larger object, a *meteorite*, survives heating in the atmosphere and lands intact. Meteorites weighing from ounces to tens of tons have been recovered.

Not until 1803 was the possibility that pieces of rock or iron might fall to earth from space regarded as fact. Skepticism against this possibility used to be so great that two centuries ago Thomas Jefferson is alleged to have said, "I would prefer to believe that two Yankee professors would lie rather than that stones could fall from heaven." How shocked he would have been to learn what else falls from the sky. The turning point in dispensing with that skepticism came when in 1803 near the town of L'Aigle in France bright flashes lit up the heavens and stones were seen to fall out of the sky over a large area.

When a meteor trail is particularly bright, sometimes so bright as to illuminate everything for miles, it is called a *fireball* (Figure 3-2). The object that causes a display like this is larger than the typical pea-sized shooting star; it is more like an apple.

In February 1994 the brightest fireball in 19 years was seen over the western Pacific. For an instant it was as bright as the sun and it was seen by many people as well as several earth-orbiting spacecraft. The technical data from the satellites suggested that the object produced an 11 kiloton explosion at an altitude of 120 kilometers.

When a fireball explodes at the end of its trail it is called a *bolide* and in rare cases people have been lucky enough to hear the sound. Such an event occurred on April 30, 1995, as reported by Peter Birch of Perth Observatory.

> At 1757 UT on 30/4/95, thousands of Perth (Western Australia) residents were awakened by the sonic boom from a meteor. Eyewitness reports indicate a track of SW to NE, and timings between sight and sound indicate a height of 15–20 km.

The object split into 4 bits around 50 km NE of Perth above the National Forest. The object was extremely bright, and short lived. No falls were observed. Reports have come from around 100 km either side of Perth.

This report was forwarded to me by a colleague using the Internet. I also received the following dramatic first-person account written by Perth high school student Alex Ringlis:

At two A.M. this morning I had the joy of just about falling asleep when I heard what literally sounded like the petrol station around the corner getting nuked, and

Figure 3-1 Artist's conception of the Leonid meteor shower as seen by Andrei Ellicott off the coast of Florida, November 12, 1799, at approximately three o'clock in the morning. From *The Midnight Sky* by Edwin Dunkin (London, c. 1870). (Courtesy D. K. Yeomans)

Figure 3-2 A remarkable image of a fireball seen on September 8, 1991, one of the few ever caught by a large telescope. The trail of light left by the meteor crossing the field of view (about 7 degrees across) began just to the right of the galaxy NGC 253 in Sculptor and the meteor exploded just as it was leaving the field of view. It was so bright that reflections inside the telescope structure produced the peculiar lighting effect on the opposite side of the image. (Courtesy David Malin, Anglo-Australian Observatory)

having a fairly strong breeze pass over my face. Another family member then asked me what that noise was, and why all the doors in the house flew open unexpectedly....Well anyway as we later learned it was a meteor entering the atmosphere over the hills in the east. News reports have claimed that the noise was actually a sonic boom caused as the meteor broke up or exploded in the atmosphere....This was no ordinary sonic boom. Never in my life have I heard one as loud as this, that also resonated as much. When I say I felt a breeze pass over my face it was literally as though a giant bellows was used over me, and the vibration that went through the house sounded like someone dropped a ten-ton concrete slab in my back yard.

Later he added this for his "information superhighway" friends:

I thank those who helped me in finding answers to the nature of the boom I heard when the [meteor] passed over. Incidentally, regarding the "wind" I felt...my best explanation for that is that my door was closed, and my window was open about an inch or two. The sonic boom must have caused some sort of shock wave through that space that I could feel? The reason that I think this could be the case is because another member in my family also reported that two of the doors in her room (one was closed) literally "popped" open with the boom.

Finally on a more humorous note....One of the local television networks is reported to have set up a "meteor-cam" wide-angle camera on their roof in the hopes of catching the big story should another large meteor decide to follow its predecessor. (Isn't this almost statistically impossible?)

Alex Ringlis was correct in estimating the statistical likelihood of a second fireball/bolide occurring over Perth in the foreseeable future. Based on what is known about the frequency of fireball events, that TV network camera may have to be kept going for several thousands of years before it would catch another event like it. Oddly, though, two years before two fireballs were seen over southeastern Australia separated by one week, both in April as well.

A fireball that seems to reach the ground usually points toward a meteorite *fall*. A meteorite *find* occurs when you trip over one in a field somewhere (Figure 3-3). If a fall is well observed, and someone takes the time to collect eyewitness reports, it is possible to make an educated guess as to where the object landed. It has become a lucrative hobby for those eager to earn added income through selling fragments of meteorites to collectors to go in search of meteorites pinpointed in this manner. Sales of space debris have recently become big business. According to Richard Norton, who has collected fascinating stories about meteorite hunting in his book, *Rocks from Space*, the going price in 1994 for common meteoritic samples was about $0.25 to $0.50 per gram whereas slivers of famous meteorites such as the Allende and the Murchison sold for $40 to $50 per gram, more expensive than gold, which at $400 per ounce converts to about $14 per gram.

Following the Perth fireball, the hunt for meteorites was on. Dr. Alex Bevan, curator of mineralogy and meteoritics of the Western Australian Museum, led the search and sent this over the internet:

> I also heard the sonic phenomena associated with the fireball. I have also interviewed more than 30 people this week....The noises were classic examples—I have heard them once before. There was a deep rumbling noise and two loud detonations (sonic booms). Those further away from the flight-path heard what sounded like one detonation. The ground vibrations caused by the aerial shock waves were so strong that they were recorded on seismographs. Those close to the flight path also heard "hissing" noises coincident with the luminous phenomena. These curious electrophonic sounds have been debated for many years but are now thought to be low-frequency radio transmissions generated by the fireball.

The cause of strange hissing sounds associated with fireballs was only recently determined. According to Colin Keay of the University of Newcastle, "For about 10 percent of those who do witness a very luminous meteor fireball, the mental

5 cm

2 in.

Figure 3-3 A 3-kilogram meteorite, this one thought to be a piece of the asteroid Vesta. The meteorite is made of material common to lava flows and its chemical signature is identical to that of the asteroid as observed from earth. It is probably a chip off the surface of Vesta and it fell to earth in 1960 in Western Australia. (Courtesy R. Kempton, New England Meteoritical Services)

impression is heightened by strange swishing, hissing, and popping noises coincident with its passage across the sky." Such sounds are anomalous because they are heard at the moment the fireball is seen, not when the sonic boom arrives. This means that the "sounds" traveled at the speed of light, a contradiction in terms. That is why for centuries they were believed to be a psychological illusion.

Keay has found that the first record of the phenomenon goes back to 817 A.D. in China. Also, Acts 2:2 in the Bible may also have been a description of bolide sounds. "And suddenly there came from heaven noise like a violent rushing wind, and it filled the whole house where they were sitting." Acts 2:3 continues with "And there appeared to them tongues as of fire distributing themselves...." This does seem to be a description of a fireball breaking up in the sky and exploding.

When the hissing phenomenon was first considered by well-known scientists, such as Edmund Halley, who in 1719 investigated such a report, it was dismissed as "the effect of pure fantasy." The secretary of the Royal Society in Britain in 1784 was also skeptical and suspected a psychological explanation. Keay points out that "These conclusions, by eminent men, bedeviled the studies of electrophonic sounds for two centuries." The bias toward a psychological explanation lasted well into the twentieth century.

After collecting reports and doing experiments in anechoic chambers using low-frequency radio waves, Keay concluded that the geophysical electrophonic sounds, as they have come to be called, are almost certainly produced when a radio wave is generated through plasma (electrical) processes in the fireball trail. Individuals hear the "sounds" only after the radio signals are detected by some means; that is, are converted from radio to sound energy in the immediate vicinity of the individual. This conversion requires what is a called a transducer that is capable of responding to very-low-frequency radio waves. Examples of effective transducers include long, loose, or "frizzy" hair, spectacle frames, and in one case a motorcycle policeman's helmet. Objects in the room around the person who hears the sounds may also act as transducers, which gives them the feeling that the sound is all around them.

At the time of writing this chapter, the hunt was on for the Perth meteorite falls, a worthy endeavor for its scientific value. Around the world and mostly during this past century several types of meteorites have been found. *Iron* meteorites make up about 4 percent of falls. They tend to be older because they have been lying about with little likelihood of crumbling to nothing, which is why they are preferentially selected in finds. They show interesting patterns in their crystalline structure, which prove that they cooled very slowly. That means they once existed inside large bodies, at least 30 to 50 kilometers in diameter, which were smashed to pieces millions of years ago. The parent bodies were almost certainly asteroids, fragments of which were then propelled into earth-crossing orbits. Most of the fragments are small, but we should not be surprised when an odd giant arrives. The dinosaurs, we must assume, were very surprised.

Then there are *stony* meteorites with a subgroup known *chondrites* that comprise 80 percent of falls. They consist of dark granular rocks containing small glassy bits called chondrules, which show they cooled quickly. These may be bits of molten rock splashed off other planets or moons. They are very old and every indication is that they were never melted as a whole.

Carbonaceous chondrites make up about 6 percent of falls. They contain volatile material, which includes water and certain molecules that easily leak out of the meteorite, like air escaping from a balloon. They are also rich in organic compounds and this argues that they were never heated very much. The carbonaceous chondrites may be the most pristine stuff left in the solar system, never having partaken in any aspect of planetary formation, otherwise the volatiles and organic compounds would long ago have been driven out.

Achondrites are stony meteorites that are very uniform in appearance that contain rounded grains that may be material stripped from the surface of some larger body such as an asteroid. The last type is known as *stony-iron*, which originally solidified in regions of molten rock and iron, probably inside planetesimals, the objects out of which the planets were formed.

In 1969 a fireball near Allende in Mexico exploded in the air and showered the area with 4 tons of fragments. About 2 tons of carbonaceous chondrite were recovered from an area about 50 kilometers by 10 kilometers. This is known as a *strewn field.* These meteorites were found to contain metals like calcium, aluminum, and titanium that solidified first in the solar nebula, the disk of hot gas and dust surrounding the protosun within which the planets were formed. The Allende is a very old sample of the solar nebula material and what is almost eerie is that this meteorite contained amino acids, both the alien and normal types.

Because meteorites fall to earth with almost monotonous regularity, it is a wonder that no one has been killed in recent times, although there are a few dramatic examples of close encounters. On November 30, 1954, a small meteorite weighing 19 kilograms struck Mrs. E. Hulitt Hodges of Sylacauga in Alabama on the thigh while she was sleeping on a couch. The object came through the roof, bounced off a radio, and gave her an ugly bruise. It must be one of life's more bizarre experiences to be struck by a rock that has traveled through space for millions of years after being created by a collision between asteroids hundreds of millions of kilometers from earth.

In recent times, automobiles seem to have been most vulnerable to meteorite attack. In 1938 in Benid, Illinois, a meteorite fell through Ed McCain's garage and then through his automobile's roof. Then, on October 9, 1992, a red Chevrolet Malibu belonging to the Knapp family in Peekskill, New York, was badly damaged by a 57-kilogram meteorite that went right through its trunk. That night many thousands of people along the east coast of the United States saw a spectacular fireball cross the sky and it is very likely that many more meteorite fragments remain to be found in that area. Richard Norton reports that the Chevy-destroying meteorite was sold for $69,000, which averages out to about $6 per gram. The automobile, too, became a collector's relic and allegedly fetched $10,000. The moral of the story is that if you manage to place your old car in the path of a meteorite it may yet fetch a nice profit. I wonder whether such manna from heaven is tax-free.

On June 21, 1994, a 2-kilogram meteorite is alleged to have smashed through the front window of the automobile driven by José Martin traveling with his wife from Madrid to Marbella in Spain. It bounced off the dashboard and landed in the back seat. Martin suffered a broken finger. Had the meteorite struck a few inches to either side, someone could have been killed. In any event, a rock from space on the forehead would not have been a happy experience. A later search of the area where the collision occurred turned up a further 200 kilograms of meteorites.

While completing the manuscript for this book, a report surfaced of yet another automobile strike, this one in Japan where Keiichi Sasatani's car suffered a dent in its hood. The incident began as a fireball seen crossing the sea of Japan that broke into four separate trails, one of which lead to an egg-sized object falling onto his car.

Based on these limited data, we conclude that meteorite falls pose a greater threat to automobiles than to people, or does it mean that there are more automobiles in the world than people? The sudden increase in the number of direct encounters between automobiles and meteorites (three in four years) does suggest that we exercise greater caution during our daily commute! Look left and right before crossing an intersection, watch your rear-view mirror, and look up in case you are about to be flattened by something left over from the formation of the solar system.

A criticism leveled at those who have recently begun to sound warnings about the threat of comets and asteroids is that no one is known to have been killed by a meteorite in recorded history. The implication is that the danger is not real. However, I should restate this criticism as follows: No one who has been killed by a meteorite in recorded history has lived to tell the tale. This is an important distinction. It is only when someone is witness to the fatality that it would be talked about.

Upon closer study, it turns out that meteorites have killed in recorded history, at least according to Kevin Yau and colleagues at the Jet Propulsion Laboratory in Pasadena, California. They examined ancient Chinese records and found more than 300 records of meteorite falls between 700 B.C. to A.D. 1920. The earliest record of a meteorite fall dates to December 24, 645 B.C., in a report attributed to Confucius. A pattern that emerged from the Chinese records was of an excess of falls from 1840 to 1880, which accords with data from European sources. Such an excess is peculiar, because there is no reason to expect the number of meteorites to vary from year to year, given that most of them have been traveling around the sun for millions of years before hitting earth. This makes it difficult to imagine why they might have remained bunched for so long.

What is striking in the Chinese records is the number of people that are said to have been killed. At least six instances involving fatalities were reported since 1300. Around 1341 several independent accounts told of a rain of iron that killed many people and animals. Something similar happened around February–March 1490 when "stones fell like rain" and 10,000 were killed. A rain of stones is not as impossible as it sounds. In 1868 near Warsaw, Poland, an estimated 180,000 meteorite fragments fell to earth in a desolate area. Although no fatalities were reported, a meteorite shower such as that in a populated area could cause considerable loss of life. (It would also make a lot of people rich from selling the fragments they might pick up in the stampede to follow!) The Chinese records told of other fatalities from falling stones that occurred in 1639, 1874, 1907, and 1915.

Yau and his colleagues interpret these data to suggest that the risk of being struck by a meteorite is far greater than has heretofore been imagined. They estimate that one fatal accident should occur somewhere on earth every 3.5 years, yet there is no evidence that the incidence is actually this high. It might be argued that most of us spend so much time indoors or concentrated in cities that we are less likely to be struck, or even hear of anyone who was been. On the other hand, I wonder whether the death of a single person by a rock from space would even be recognized as such. Perhaps meteorite impacts can account for some cases in police files of unsolved "murders" where the victim was struck on the head by an unknown assailant. Someone might like to do a research project on this, although I hope this does not become a defense for real crimes.

From a study of fireball paths, meteorites appear to arrive at earth along orbits that originate in the asteroid belt. They are therefore regarded as fragments of asteroids that collided and shattered, in many cases millions of years ago, as inferred from chemical analysis that reveals when they were last part of a larger object. Their absolute ages, reckoned from radioactive dating, shows that they were part of their parent bodies 4.6 billion years ago.

The origin of meteorites was not always well understood. In 1893 Sir Richard Gregory, a professor of astronomy at Queen's College in London published a book entitled *The Vault of the Heavens* in which he wondered about their origin. He reported that astronomers believed that meteorites were portions of matter ejected from volcanoes on earth, the moon, and the planets. Those were supposed to have blasted debris into space, which then escaped the gravitational trap of their home bodies. This suggested that some of the meteorites that fell to earth might have come from another planet. Oddly enough, this happens to be the explanation for some of the 14,000 meteorites found in a limited area of Antarctica. A few of them appear to be Martian rocks, and some of lunar origin have also been found. This means that moon rocks were lying about on the ice sheet all along, even as Project Apollo sent men to the moon in 1969 to bring back samples. (Think of the money that could have been saved.) However, contrary to the idea that held sway the past century, the Martian and lunar meteorites were not blown into space by volcanoes. Instead, they were splashed into space by impacts with asteroids.

Other objects littering the earth tell of past impacts. These are *tektites*, black, glassy objects typically of the size of one's thumb, which were formed when molten material was splashed into space after an impact of a giant meteorite or an asteroid with the earth. While traveling through space they cooled and turned to glass that took on interesting shapes, many of them buttonlike, the hallmark of tektites. There are enormous tektite-strewn fields spread over various parts of the world. The most famous are in Australia, Vietnam, and Indonesia, deposited about 700,000 years ago.

Sometimes quite large objects fall to earth, such as the one that created the famous Barringer meteorite crater in Arizona, 1.2 kilometers across, 200 meters deep, and 49,000 years old (Figure 3-4). The impactor was an iron meteorite about 40 meters wide according to one estimate. Craters are, in general, expected to be about 20 times wider than the object that slams into the ground. The impact would have produced something like a 50-megaton blast, an explosion that today would wipe out a large city and its suburbs in an instant.

Figure 3-4 The Barringer meteor crater in Arizona, 1.2 kilometers wide and nearly 200 meters deep. The impact that created this crater was in the 100-megaton range, capable of devastating a wide area of countryside out to 80 kilometers from the blast. (Courtesy Richard Grieve, Geological Survey of Canada)

About 150 impact craters (see chapter 12) have now joined the list of which the Barringer was once the only certain member. Most of them were formed by objects too large to be called meteorites anymore. The larger impactors were probably members of the *asteroid* family, chunks of rock from meters to hundreds of kilometers across that wander about the sun between the orbits of Mars and Jupiter (Figure 3-5). When the solar system formed, Jupiter's gravitational influence produced huge tides that prevented a planet from forming in the region of space that now contains the asteroid belt.

The first asteroid was discovered January 1, 1801, by Guiseppi Piazzi. Called Ceres, it is 1000 kilometers in diameter and it was found because of a rather odd coincidence concerning planetary orbits. In 1766 two German astronomers, Titius and Bode, noticed that there was a large gap between Mars and Jupiter, which might well include a planet. So the search began. Instead of a large planet, astronomers found a lot of little ones—minor planets as they are sometimes called. The next four, Pallas, Vesta, Juno, and Astraea, were discovered by 1845 and then the pace of discovery quickened. During the next 150 years a further 20,000 main-belt objects were detected. The luxury of giving them individual names is no longer applied to the smaller ones. Instead, a date and number is used, with a number assigned when an orbit for the object is known. Over 3500 main-belt asteroids have good orbit estimates and another 6000 have poorly determined orbits. The others have not been seen often enough to derive orbits.

Over half a million main-belt asteroids with diameters greater than one kilometer probably exist. They swarm around the sun between the orbits of Mars and Jupiter in a vast region of space about 2 astronomical units (AU) wide and 6 AU thick. (An astronomical unit is 150 million kilometers or 93 million miles.)

The relative absence of smaller, free-floating objects between the larger asteroids is attested by the success of flying five deep-space probes, including two *Pioneers*, two *Voyagers*, and *Galileo*, unhindered through the main belt. Collision with even a small object no larger than a pea could have destroyed any of these rapidly traveling spacecraft.

There are two groups of asteroids in Jupiter's orbit known as the Trojans that follow and precede Jupiter in its orbit and are separated from the planet by 60° of angle as seen from the sun. This is the result of a peculiar gravitational phenomenon predicted by the mathematician Lagrange during the late eighteenth century.

Distinct gaps in asteroid distances from the sun are known as Kirkwood gaps. These are due to gravitational resonances with Jupiter, which means that Jupiter gives any asteroid that finds itself in the gap a regular tug every once in a

Figure 3-5 The asteroid 243 Ida and its satellite, Dactyl, photographed by the *Galileo* spacecraft on its way to Jupiter in 1994. Ida is about 56 kilometers long and the photo was taken from a range of 10,800 kilometers. The asteroid, which was the 243rd to be discovered in the asteroid belt, is heavily cratered. Dactyl is only about 1.6 kilometers across and appears to be orbiting Ida at a distance of about 100 kilometers. (Courtesy NASA)

while, which causes it to move away from that region of space until, after millions of years, the gaps are cleansed of objects.

It is broadly accepted that the main-belt asteroids are fragments of planetesimals (larger lumps of matter) left over after planetary formation, but some of them may be remnants of comets. There is an asteroidal object named 2060 Chiron (pronounced Ky-ron) looming between the orbits of Saturn and Uranus. It is a lone wanderer about 180 kilometers in diameter. Following its discovery in the 1970s it was thought to be an asteroid. Then, in the 1990s, it began to exhibit a cometlike appearance, to which we'll return in the next chapter. Chiron is one of those "in between" objects—both comet and asteroid, depending on when you look at it.

Asteroids come in several types, classified according to how they look from earth. They reflect different amounts of sunlight and may have originated in different regions of the asteroid belt, which suggest different histories.

This brings us to the potentially dangerous asteroids, those whose orbits cross that of the earth. Long regarded as an interesting anomaly, the NEAs (for near-earth asteroids) have suddenly, and for very practical reasons related to the continued survival of our civilization, become the focus of intense study.

The first earth-crossing asteroid, 887 Alinda, was discovered in 1918 but it was not recognized to be in this category until a close study of its orbit was carried out in 1970. In 1932 two NEAs were discovered that became prototypes for different populations of such objects. The first, 1862 Apollo, is 0.7 kilometer across and orbits the sun in 1.78 years. Its name was subsequently given to the class of objects that actually cross the earth's orbit and spend most of their time just beyond it.

The second, 1221 Amor, came to represent the prototype of a group that cross Mars's orbit and come close to the earth, sometimes crossing its orbit as well. Many Amors that do not cross the earth's orbit at this time are known, but they remain potentially dangerous because they can be unpredictably redirected by Jupiter's gravitational influence to become earth crossers. The problem is that predictions involving the gravitational influence of more than two masses, in this case the asteroid, the sun, and Jupiter, are notoriously difficult to make, and usually inaccurate when calculated more than a few decades into the future. One potentially lethal Amor that does not currently cross the earth's orbit is 433 Eros, a 20-kilometer-diameter object that has the potential to collide with the earth sometime during the next 400 million years. It was an object of this size that wiped out the dinosaurs and Eros is a sobering reminder that near-earth space even now contains such behemoths.

A third class of earth crossers are the Aten asteroids; these spend most of their time inside earth's orbit. The first one, 2062 Aten, was discovered by Eleanor Helin in 1976 in her continuing quest of NEAs using a small telescope on Palomar mountain.

By 1982 a total of 49 earth crossers had been discovered; 4 Atens, 30 Apollos, and 15 earth crossing Amors. It was then estimated by Eugene Shoemaker of the U.S. Geological Survey in Flagstaff that as many as 1,800 remained to be discovered. With the recognition that these earth-crossing asteroids pose a threat to our planet, searches were stepped up and by 1992 another 110 had been found. The total number rose to over 350 in 1995.

The largest of the current earth crossers are 6 to 8 kilometers across and go by the names 1627 Ivar and 1580 Betula. Most of the others so far discovered are in the kilometer-size range, lethal for civilization if they should ever slam into the earth. Following the spate of discoveries in the early 1990s, it has been estimated that between 5000 and 10,000 earth crossers of more than half a kilometer in size may exist. Collision with an object this large is likely to bring civilization close to the brink, even without wiping out species.

The origin of the various earth-crossing populations is not at all clear. Unlike the main-belt asteroids, they cannot have been around since the formation of the solar system. In order for them to have survived planetary collisions as well as gravitational perturbations ejecting them from their earth-crossing orbits, the initial population would have had to include more mass than is in the sun. These objects must, instead, have been injected into earth-crossing orbits well after the earth was formed. Some of the earth crossers may have come from the asteroid belt after being perturbed by the gravitational pull of Jupiter, or by collisions within the asteroid belt, but many of them may be the burned-out hulks of old comets that broke up and established orbits that now haunt the earth.

On October 11, 1983, John Davies and Simon Green at the University of Leicester in England found an asteroid in the data obtained by the Infrared Astronomical Satellite. It is an earth-crossing Apollo and was given the name Phaeton. It orbits the sun in a mere 1.4 years and it turned out to be associated with the Geminid Meteor Stream, which produces a meteor shower every December (see Table 3-1). Since meteor showers are associated with comet debris, this NEA is probably a comet hulk.

Some of the stuff in earth-crossing orbits may have come from the moon. From the discovery of moon rocks in Antarctica, it has been estimated that 80 to 90 percent of material blasted off the moon by impacts goes into orbit around the sun. Some of it returns to the moon or earth within 10 million years and other bits may even fall onto Venus or Mars.

According to Paolo Farinella of the University of Pisa in Italy, who together with a team of colleagues uses computers to follow asteroid orbits, many NEAs will inevitably fall into the sun. They followed the paths of 47 NEAs in a computer and found that in 2.5 million years 15 of the objects will strike the sun and 4 will be ejected from the solar system. Comet Encke, which is in an earth-crossing orbit, will fall into the sun in about 90,000 years time. Such calculations confirm that NEAs cannot remain in their present orbits for many millions of

TABLE 3–1 Meteor Shower Timetable

Shower name	Date of maximum	Associated comet
Quadrantids	January 4	—
Lyrids	April 21–22	Thatcher (1861 I)
Pi-Puppids	April 24–25	Grigg-Skjellerup
Beta Aquarids	May 3–5	Halley
June Draconids	June 30	Pons-Winnecke
Beta Taurids (daytime)	June 30	Encke
Delta Aquarids	July 29–30	—
Aplpha-Carpicornids	August 2	Honda-Mrkos-Pajdusakova
Perseids	August 11–12	Swift-Tuttle
October Draconids	October 8–10	Giacobini-Zinner
Orionids	October 20–21	Halley
Andromedids	November 1	Biela (extinct) and shower nearly extinct
Taurids	November 3–4	Encke
Leonids	November 16–17	Temple-Tuttle
Geminids	December 13–14	Phaeton (asteroid)
Ursids	December 22	Mechain-Tuttle

Note: For the time and date for the peak of any shower in a given year, consult any monthly astronomy magazine that usually advises its readers where and when a shower peak is expected.

years, which means that they must have been injected relatively recently, as seen from an astronomical perspective.

In 1993 the *Galileo* spacecraft en route to Jupiter flew by and photographed 243 Ida, which, it turned out, has a moon, since named Dactyl (Figure 3-5). Both Ida and Dactyl are cratered, and Ida is so covered in craters that it must be at least a billion years old. However, orbital calculations show that they cannot have danced around each other for that long; Richard P. Binzel of MIT thinks that within 100 million years such objects would be blasted to smithereens by collisions with other asteroids, so why are Ida and Dactyl still locked in an embrace? No one yet knows.

During the last few years other asteroids have also been seen up close and many of them are paired. Some have been mapped with powerful radar transmitters capable of bouncing radio waves off the distant rocky lumps. Planetary scientists like Steve Ostro of NASA's Jet Propulsion Laboratory are capable of interpreting the returned echoes to make a map of an asteroid. In 1989 Ostro's radar data showed that 4769 Castalia (Figure 3-6) was double, probably a contact binary, which means two objects in contact but separate. Timing of the radar echoes also allows an asteroid's location, shape (Figure 3-7), and motion to be measured very precisely, which means that its orbit can be perfectly determined, a useful talent when it comes to predicting the likelihood that a given NEA will strike the earth.

Several other NEAs are doubles according to radar data, which show them to have a dumbbell shape, and others are contact binaries. This is important to the understanding of what may have happened on earth 65 million years ago, when an impact triggered the mass extinction event. Iridium data from the United States show that there may have been at least two major blasts due to multiple impacts.

Oddly enough, the possibility that asteroids might be double was first noticed by amateur astronomers observing occultations (eclipses) of stars by asteroids. For years they had reported occasional "wink outs" before the asteroid moved in front of the star, which suggested companion objects close to the asteroids. Few professional astronomers believed them then, but now it seems those amateurs may have been sharp-eyed enough to have been correct.

The NEAs discovered to date are mostly larger than 100 meters across, but in 1993 the possible existence of a subgroup of asteroids was reported. David Rabinowitz and colleagues at the Lunar and Planetary Laboratory of the University of Arizona in Tucson are involved in a continual search for NEAs as part of Project Spacewatch. In two years after their search started in earnest (on January 1, 1991), they found 40 NEAs, including 13 with diameters smaller than 50

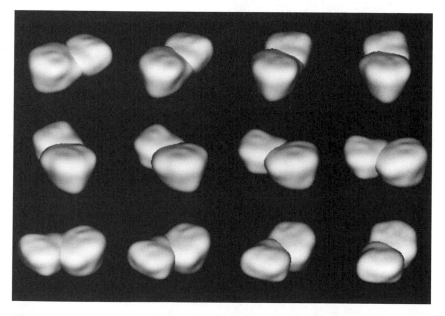

Figure 3-6 Computer-generated model of 4769 Castalia based on radar imaging data created by Scott Hudson of Washington State University and Steven Ostro of the Jet Propulsion Laboratory. The object, a contact binary, is a little over 100 meters in diameter. The radar data on which these images are based were obtained by Arecibo radio telescope when Castalia was 5.6 million kilometers from earth. (Courtesy Steve Ostro and JPL/NASA)

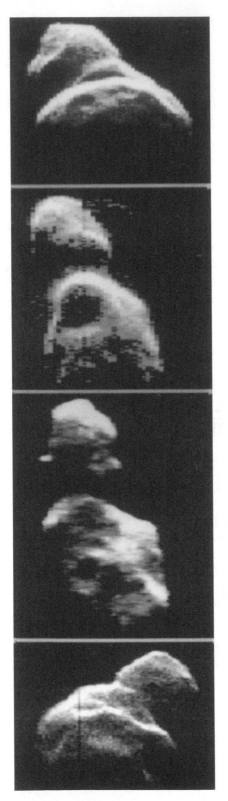

Figure 3-7 Radar images of the asteroid 4179 Toutatis made using the Goldstone radio telescope in California in December 1992 when the asteroid was at its closest to earth. The asteroid is a double with the two fragments about 2.4 and 1.8 kilometers across. The images were made over a period of five days and show the asteroid rotating. The large crater visible in the second from the top image is about 650 meters across. (Courtesy Steve Ostro and JPL/NASA)

meters. These have orbits similar to that of earth and may form a local asteroid belt not previously detected. They think that it is difficult to account for these small earth-approachers (SEAs) unless they were ejected from the moon as a result of impacts with larger objects. From 1991 to 1994 three SEAs passed closer to the earth than the moon's distance (see chapter 9).

The existence of thousands of NEAs poses an extraordinary threat to our continued existence, certainly on time scales of millennia, but they are not the only objects capable of crashing into the earth to produce devastating consequences. Comets that wander close to the sun after spending billions of years in space beyond the solar system pose another hazard.

4

SOMETHING ABOUT COMETS

DISCOVERY of the iridium layer in the K/T boundary clay was the first clue that pointed to a cosmic impact as the trigger of the mass extinction that wiped out the dinosaurs and their associates. But what hit the earth? Was it an asteroid or a comet? To answer, we need to know something about their differences. Unfortunately, the distinction is very blurred.

Comets are thought to be huge icy objects, probably with cores made of a mix of water ice and silicates (sandy material), pristine examples of the type of material out of which the solar system was formed. Some of them are hundreds of kilometers in size and they may have been built in the envelopes of gas and dust that surround cool, supergiant stars at the end of their lives. Part of the doubt about distinctions comes from trying to decide what a comet would be like after the ice evaporates. Would it then be like an asteroid?

Around the end of the nineteenth century the British astronomer Sir Richard Gregory pictured comets as made up of a cloud of meteorites. He thought that when such an object was first pulled into the solar system from interstellar space it began to glow because of internal heat created as particles began to jostle one another. As the object drew closer to the sun a tail was supposed to be formed as the particles between the meteorites bumped into one another and began to escape.

He did consider the potential risk to earth if it were to run into the head of a comet made up of lots of meteorites. The picture he painted was based on what an earlier astronomer, Sir Simon Newcomb, had written about this possibility. Newcomb admitted that, although there were more likely ways to die than as a result of comet collision, such a fate was real. Should such a collision, occur, Gregory conjured up a picture of what might happen. On the one hand, if the comet head was made up of dust, the earth's inhabitants would experience nothing more than a stunning display of shooting stars. But if the comet head was made of cannonball sized objects the consequences would be dire.

> Myriads of meteoritic masses would beat upon the earth, and the burning of the materials of which they are composed would probably use up the oxygen in the atmosphere, in which case, man and all the animal creation would perish. The temperature of the air would be raised to such a degree that all vegetation would be destroyed and our world would be transformed into a desolate and barren rock.

He followed his description with a reminder that the prospect was not pleasant, a stunning piece of understatement, and reassured his readers that this fate would not be expected for another twenty million years.

His imaginary view of the fate of earth after a comet impact has been vastly improved by modern computer simulations as regards the nature of its consequences. The timing of the next impact has also been looked at a lot more closely, as we shall see in chapter 13.

Comet visits, which are called apparitions, have been noted by curious astronomers for over 3000 years, but it wasn't until the seventeenth century that it was appreciated how many of them make regular close passes by the sun. Because comets follow mostly highly stable orbits, the predictability of their return makes them seem almost friendly. That is not how ancient peoples greeted their arrival, however. A new comet was almost invariably cause for concern, if not downright terror.

The most famous of the regular cometary visitors is Halley's comet (Figure 4-1). Its orbit was determined by Sir Edmund Halley back in 1695, using the laws of gravity that had been newly discovered by his friend, Sir Isaac Newton. Halley found that this comet reappeared every 76 years and the next visit would occur in late 1758 or early 1759, by which time he was, unfortunately, dead. It befell the great comet hunter, Charles Messier, to spot the comet in January 21, 1759, but unfortunately for him he did not get the credit.

Messier had been searching for Halley's comet for 18 months, the last two of which were dogged by nearly continuous cloud cover over France, which prevented him from looking at all. Within weeks of his spotting the comet it moved behind the sun and was not seen again until it reappeared on the other side a few months later. At the time Messier was an assistant to Joseph-Nicholas Delisle who, as his boss, was able to dictate that Messier should not reveal his discovery,

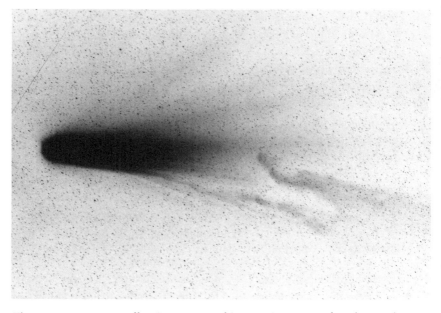

Figure 4-1 Comet Halley in 1986. In this negative print of a photo taken on March 10 of that year, a segment of the comet's tail had detached itself and has separated from the comet head by 7.3 million kilometers. The photograph was taken by the 1-meter Schmidt telescope of the European Observatory in La Silla Chile. (Courtesy European Southern Observatory)

perhaps because Delisle, too, had been searching. They must wait until it reappeared, Delisle ordered. As it turned out, the announcement was delayed for two months, by which time it was too late.

When Halley's comet reappeared on April 1, 1759, Messier and Delisle were stunned to learn that a few weeks before Messier's initial success it had been seen by an amateur astronomer, Johann Georg Palitsch, in Germany. He found it on Christmas Eve, 1758, and did not realize it was Halley's comet. Announcement of Palitsch's discovery made its tortuous way into the world via a Dresden newspaper and finally reached Paris. Donald Yeomans has written about the subsequent furor, "The collective pride of the Paris academy must have been deeply wounded, since the recovery of comet Halley was made by a German amateur more than three months before they [the French] knew anything about it."

The successful prediction of comet Halley's 1758–59 return was a dramatic confirmation of the predictive power of the physics of moving objects (dynamics), and heralded a new era in science in which the cause of the motions of heavenly bodies became understood.

Messier grew famous in his own right for his catalog of fuzzy cloudlike objects among the stars that he said should be avoided by comet hunters. Some of them turned out to be star-forming nebulae in the Milky Way and others were distant galaxies. The objects from his list of "things to be ignored" still bear their Messier catalog numbers—for example, M31, the great galaxy in Andromeda.

When Halley's comet visited in 1910 (Figure 4-2), it arrived to considerable fanfare and great consternation. The earth was expected to move through its tail and the prospect seemed to awaken deep societal fears of comets in general. Millions awaited the event with awe and consternation in their hearts. When the earth passed through the tail, people panicked. Greatly increased meteor activity was seen, and by then word had gotten out about the discovery of cyanide in the comet. Cyanide is a terrible poison when taken in terrestrial doses, as readers of classical murder mysteries can attest, but in comets this organic molecule is present in tiny concentrations that could never harm anyone on earth, even if the planet were bathed in the comet's tail.

A typical comet is made of at least three distinct parts; a *coma*, the glowing head created by gas and dust evaporated by sunlight from the comet's *nucleus*. This, in turn, consists largely of water ice polluted by dust and complex organic molecules; an icy dirtball some have called it, a term that has superseded the dirty snowball label. The nucleus may be built around a core that is a mix of water ice and silicates, although the close-up look obtained by the European spacecraft, *Giotto*, of Halley's comet (Figure 4-3) brings to mind an image that

Figure 4-2 Comet Halley as photographed in 1910. Great fear existed that, when earth passed through Halley's tail, widespread pestilence would result. It didn't. (Courtesy Yerkes Observatory, University of Chicago)

Figure 4-3 The nucleus of comet Halley as photographed by the Halley Multicolor Camera on board the European spacecraft, *Giotto*, in 1986. Close examination of the *Giotto* photographs allowed structure on the comet's surface to be identified, including the presence of a crater. The bright areas indicate where gas and dust are escaping the nucleus, material that will contribute to its tails. (Courtesy H. U. Keller and Max Planck Institute for Aeronomy)

the core may be more solid. A gas *tail* can stretch 10 million to 100 million kilometers into space away from the comet *head*, the term used to describe the coma and nucleus together. The tail is blown away by solar wind and glows because the gas is hot enough (many thousands of degrees Celsius) to be ionized. A second tail is created from dust left in the wake of the comet (Figure 4-1).

Sometimes tails disconnect to form new ones, especially at what are called sector boundaries, regions in space around the sun where the solar magnetic field direction suddenly reverses. Those magnetic fields, in turn, are rooted in the solar surface and flail through interplanetary space making patterns like a rotating sprinkler sending out streams of water in spiraling swaths over a lawn. Watching to see how comet tails change from day to day tells planetary scientists a great deal about conditions in the wind of particles and magnetic fields that blow out of the sun.

For some reason, comets have always been regarded as the portents of doom, which raises interesting questions being asked for the first time. Is it possible that human societies carry a collective memory of a relationship between comets and catastrophes? Why else would people be afraid of them? If comets never harmed anyone, why would comet fear be so deeply etched in the legends and sagas of many cultures?

A telling description of people's reactions to a relatively close comet encounter is the following quotation from an Atlanta (Georgia) newspaper report related to the November 1892 close pass of Biela's comet (quoted by Herbert Howe in 1897):

> The fear which took possession of many citizens has not yet abated. The general expectation hereabouts was that the comet would be heard from on Saturday night. As one result, the confessionals of the two Catholic churches here were crowded yesterday evening. As the night advanced there were many who insisted that they could detect changes in the atmosphere. The air, they said, was stifling. It was wonderful to see how many persons gathered from different sections of the city around the newspaper offices with substantially the same statement. As a consequence, many families of the better class kept watch all night, in order that if the worst came they might be awake to meet it. The orgies around the colored churches would be laughable, were it not for the seriousness with which the worshippers take the matter. Tonight (Saturday) they are all full, and sermons suited to the terrible occasion are being delivered.

Imagine all those people unloading a litany of their sins in a great community-wide catharsis. I am curious about how those people felt in subsequent weeks, after the scare was shown to have been unfounded.

Communal reaction to pending comet flybys is not a thing of the distant past. In recent times, the visit of comet Kohoutek in 1973 created a considerable flurry of interest, anticipation, and apprehension. Astronomers gave advance warning of more than a year of Kohoutek's approach and soon the newspapers began to herald its visit in terms of superlatives. It would become the greatest cometary display ever, it was said. Before you knew it, people in all walks of life were celebrating the coming of the comet. They were not only talking but doing something about it. No matter how often astronomers went on record to plead doubt about how bright Kohoutek might become, stressing that it might yet be not much to look at after all, the imaginative cat was out of the media bag. Everyone talked about comet Kohoutek.

Rituals began to be played out to mark its coming. I personally took part in an interesting evening in a residential garage in Boulder, Colorado, where two rock bands got together to make spontaneous music. The participants began the event by bathing in the feeling of the comet. They let their personalities be taken over with thoughts of the comet's visit. A local "shaman," a university student who dabbled in esoteric affairs, painted a verbal image in mystical terms about the meaning of comets. I added words about the vast journey the comet had made, knowing that when it last visited our forebears had not even moved into the caves; Kohoutek's orbital period had been estimated to be about 80,000 years.

When all the musicians were filled with the right spirit, they began to make music, each creating his own expression, independent of the others until, for truly heavenly moments, their sounds blended. They created New Age music before New Age music was born. Occasionally the harmony was lost and periods of terrible discord rent the air. Then the players would again meld and sounds became hauntingly beautiful. The comet was on its way! Those musicians succeeded in filling their space with a musical sense of awe about the visit of Kohoutek. Theirs was a late-twentieth-century version of the furor caused in 1910 when comet Halley brought out the confidence tricksters, peddlers of snake oil, and mystics proclaiming the end of the world. Some saw Kohoutek as heralding the end of the world, hardly likely since it was not on an earth-crossing orbit. Since it was being used to support whatever cause anyone wished to sell to a gullible public, any connection with reality hardly seemed to matter any more.

As it turned out, Comet Kohoutek was a public relations failure. No one I knew personally ever actually got to see it. This later made me wonder why it had caused so much interest. Very soon after their early enthusiasm, astronomers had gone out of their way to warn that the comet might be a bust, but their voices went unheard. The Kohoutek phenomenon had a life of its own, and I think the reason was that it provided a glorious distraction from the realities of the time, from the Vietnam War trauma being experienced in the United States, and by the malaise produced when President Nixon's gang were caught with their fingers in the doors of the Watergate complex. In 1973, as comet Kohoutek roared toward the sun from the depths of space, we shared a communal unhappiness about such matters, and the arrival of a cosmic visitor offered a diversion with almost metaphysical overtones. We pinned our hopes on it. Perhaps we wished that after its arrival things would go better. Unfortunately we were disappointed by the visit, but for a while it succeeded in offering a welcome distraction from reality.

A few years later in 1976 I saw my first and only "real" comet. No one had warned me that it would be visible, and I was an astronomer! Comet West was visible one morning around daybreak, a magnificent spectacle covering a good fraction of the eastern sky, its enormous tail shaped like a fan spread out above its head. I noticed it out of the window of my dining room. The next day it was gone. It didn't even get a mention in the local newspapers. It was as if it had never happened. Comet West was found to have been in a 16,000 year orbit but it has since been thrown out of the solar system by its close approach to the sun in 1976.

What was the difference between comet Kohoutek compared with comet West? The former was a dud heralded with great fanfare, while the latter was a spectacular visitor completely ignored. Perhaps it was because the advance warning about West's coming had been insufficient to stimulate media interest. I think not. I think that it was because comet West made its appearance in an average year, when our national psyche had settled down again. Nixon was gone, the Vietnam War was over, and we were getting on with our lives. We were no longer in need of a heavenly distraction from earthly affairs.

Comet Halley's return in 1986 was heralded with a lot of fanfare as well, but it was mostly driven by the enormous international effort on the part of astronomers to study the comet, including the visit of several spacecraft to its vicinity. Also, comet Halley was the object that had shown that the law of gravity could be understood by mere mortals. Comet Halley returned on schedule and in that sense did not disappoint, but unfortunately it did not light up the skies as some had hoped. It passed the sun too far from the earth for anyone to see it, except those who made a special effort. I made such an effort and all I saw was a fuzzy blob among the stars, barely brighter than the faintest stars visible to the naked eye. But at least Halley vindicated the astronomer's trust in the laws of nature, and it behaved in an orderly manner, which is more than can be said for the earth-crossing comet Swift-Tuttle, to be discussed in chapter 9.

Comets are found in several clouds around the sun. One is the reservoir for short-period comets such as Halley that pay regular visits every hundred years or so. They spend most of their time in a region of space beyond Neptune in what is called the Kuiper belt. There are 100 or so of these short-period comets known from studies of their orbits.

The Kuiper belt may be made of objects that never formed a planet beyond Neptune. Pluto and its satellite Charon wander about out there as well and Pluto may even be regarded as the "King of the Kuiper belt." It may not be a real planet after all, just the largest (with a diameter of 2,400 kilometers, twice as large as Charon) of many objects that have been lured out of the confines of the Kuiper belt. It is even suspected that Triton, a large moon of Neptune may be a recently captured Kuiper belt object.

The first of the real Kuiper belt objects discovered (known as 1992 QB$_1$) was observed by David Jewitt of the University of Hawaii and Jane Luu of the University of California, Berkeley. At about 250 kilometers across, derived from its brightness, or 40 times the size of a typical comet nucleus, it is not obviously a comet at all. At least 23 of these objects have been discovered in the last few years; the largest are 1994 VK$_8$ and 1995 DC$_2$, which are about 360 kilometers across. The Kuiper belt may be made of planetesimals, or the rocky, asteroid-like objects, of which the solar system formed. But even this may be a simplistic view, because of the difficulty in telling a comet from an asteroid at such huge distances.

At the time of writing it was reported that there might be as many as 10 billion comets or asteroids in the Kuiper belt of which 35,000 might be larger than 100 kilometers. This is an awesome reservoir of objects that would have the power to devastate our entire planet, should they ever enter the inner solar system. Also, the Hubble Space Telescope was used in a search of a small area and found indirect evidence for 50 to 60 Kuiper belt objects in one of its exposures.

Two of the most prominent stragglers from the Kuiper belt region of space are 2060 Chiron, whose orbit takes it inside that of Saturn and then out to Uranus, which means that is has escaped the belt and is probably headed toward the inner solar system in the long run, and 5145 Pholus, also discovered in 1992, which currently ranges from inside Saturn's orbit to just beyond Neptune.

According to Paul Weissman of the Jet Propulsion Laboratory in California, these planet-crossing orbits are not stable. On time scales of a few million years they will be ejected from the solar system, or they may spiral closer to the sun, possibly ending up in dangerous earth-crossing orbits. Their finite lifetimes suggest that they could not have been here since the formation of the solar system, so where did these objects come from? It is possible that they arrived from a cloud of comets that exists even further out, the so-called Oort cloud that is supposed to surround the sun out to a good fraction (a third) of the distance to the nearest stars. Objects out there take up to millions of years to orbit the sun. Many of these Oort cloud comets approach from random directions because they originate so far away that they scarcely heed the gravitational influence of Jupiter and Saturn, although when they wander into their vicinity those planets can redirect the comet paths in significant and unpredictable ways. It is suspected that when the sun and the Oort cloud pass through the disk of the Milky Way galaxy, comet orbits may be perturbed to send objects into the Kuiper belt where they hang about before being sent into planet-crossing orbits by the influence of the planets themselves, or by a further gentle push of passing stars.

The Oort cloud may contain a trillion comets that accompany the sun in its perambulations around the Milky Way galaxy. There must be similar clouds around other stars, which implies that there must be trillions of times as many comets in the Galaxy as there are stars. But comets are so tiny compared with the stars that their total mass is an insignificant fraction of that of the Milky Way galaxy. What is not insignificant, however, is the influence that an impact of even a small comet would have on the atmosphere of a planet such as earth.

As comet Halley moved across the skies in 1986, astronomers and paleontologists were beginning to mutter about a new danger, something very much more terrifying than a solitary comet passing the sun. Part of the danger is that comets sometimes behave in unpredictable ways. For example, some suffer from hiccups; they emit bursts of gas when quite far from the sun, which act like jets to alter the comet's orbit. Comet Schwassmann-Wachmann—a name that hardly rolls off one's tongue—misbehaves in this manner. It circles the sun between the orbits of Jupiter and Saturn where it might be expected to lead a relatively sedentary life. But this is not so. From time-to-time comet Schwassmann-Wachmann emits bursts of particles—gas and dust—that create a coma. Halley's comet, too, when way out in the solar system after its recent visit to the sun, was seen to undergo a brightening, which, it is suspected, was also related to the sudden escape of volatile material. But why?

There has been speculation that when these comets are relatively close to the sun, pockets of ice inside the comet are heated so that at some critical juncture they evaporate to produce water vapor that bursts out of the comet as it hurtles through space. Or so it was suspected until 1994. Then Matthew Senay of the Institute of Astronomy in Hawaii together with David Jewitt reported evidence for the presence of carbon monoxide in the Schwassmann-Wachmann coma. Carbon monoxide is a common molecule in interstellar space and no one is surprised that it exists in the frozen state inside comets. What is odd is that there is enough sunlight beyond the orbit of Jupiter to heat comet Schwassmann-Wachmann sufficiently for an occasional burst of gas to escape from its frozen interior.

Sir Fred Hoyle, the famous British astrophysicist thinks otherwise. He suggests that this phenomenon of comets producing outbursts of stuff to form temporary comas is due to the accumulation of by-products of metabolism of bacteria locked deep inside the comets. Occasionally the pressure of the gases they produce grows so strong that an outburst results. It is an elegant idea, but not within the mainstream of thought of cometary astronomers, so they ignore his suggestion completely. I find his theory intriguing, however. After all, why would ices inside the comet suddenly evaporate into space so far from the sun. Carbon monoxide, for example, freezes, and hence thaws, at -199 degrees Celcius. If it bursts out from beneath the skin of the comet, it means that it was warmed down there, which suggests that deeper within the comet it might be even warmer still. That warmth does not come from the sun, says Sir Fred. Perhaps bacteria are alive inside the comet and the heat is produced by their metabolism.

Overall, a comet's life is not a happy one. They are not destined to last forever, and in recent years some comets have been seen to disappear into the sun. On August 30, 1979, *Solwind*, a Naval Research Laboratory satellite, photographed a small comet just before it crashed into the sun. *Solwind 1* produced a spectacular show as it vaporized. Its remains caused a brightening over half the solar disk which lasted a full day. The spacecraft observed five more such comets, all members of a family known as the Kreutz sun grazers, although they did very little grazing while *Solwind* was watching. None of those comets survived an encounter with our star.

In 1984 a camera on board the *Solar Maximum Mission (SMM)* spacecraft also spotted *Solwind 5*, and then took over the role of discoverer. It proceeded to photograph, all by chance, ten more sun grazers in the next five years. Bob MacQueen and Chris St. Cyr, at the time at the High Altitude Observatory in Boul-

der, described these sun-grazing comets. Their data were useful to Brian Marsden
of the Harvard-Smithsonian Center for Astrophysics in Cambridge, Massachusetts, who traced the lineage of the spate of *Solwind* and *SMM* comets all the way
back to the Stone Age.

These sun-grazing comets are named after Heinrich Kreutz who, in 1891,
published a study of half a dozen of these objects, which are almost impossible to
see from the earth. Between May and October they approach from behind the
sun and then retreat in the same general direction. Their small size makes most
of them invisible except during a total eclipse, which was how the original sun-
grazing comet was found in 1882.

The Kreutz objects are examples of long-period comets that wander here
from beyond the outskirts of the solar system, which contrasts with the short-
period comets that are well-behaved members of the solar system, reflecting the
orderly motion expected of matter associated with the formation of the planets.

Marsden's family tree for the Kreutz sun grazers allows our imagination to
wander. It also allows us to appreciate a crucial aspect of comet behavior, especially that of giant comets. They can break up on passing too close to the sun,
which means where once a lone giant approached, a swarm of dwarfs retreats
after the encounter. If the original object was in an earth-crossing orbit, with little chance of earth ever running into the lone visitor, the danger of collision is
greatly enhanced after the giant breaks into hundreds of fragments.

The Kreutz family ancestor was first seen about 15,000 years ago, give or
take several thousand. Back then more than a quarter of the world was covered
in glaciers and in Europe Neanderthals had been overwhelmed by *Homo sapiens.*
In some of their caves, Upper Paleolithic people painted magnificent murals,
especially at Lascaux in the Dordogne region of France.

Those people probably saw a great comet that filled their hearts with fear
and awe. It was a natural wonder they could not comprehend, and for which a
cave painting provided no succor. It is possible that they saw it break up into several fragments, none of which happened to come anywhere near earth, which
was lucky for them, and for us. But break up it did. Some of the comets spawned
by the demise of the giant visited the sun again and again, perhaps as many as 10
to 20 times over the next 15,000 years. At each visit, some pieces burned up in
the sun while others struggled back into the depths of space, uselessly striving to
defy the pull of the sun, which would inevitably lure them back for another close
encounter, until they were all burned up.

And so it went until one of the fragments filled the night sky with a glorious spectacle in 371 B.C., the year Aristotle and Ephorus gazed at this cosmic visitor, a mere shadow of its long forgotten ancestor. That event drew particular
attention because it was associated with great earthquakes and tsunamis in
Greece. This comet then had its own close encounter with the sun, which it survived as two or three fragments that returned around A.D. 1100.

The offspring of these comets are still in perilous orbits that bring them

Figure 4-4 The comet of 1744 after it broke into six fragments. Before it reached the orbit of Mars no tail was seen. In March of that year it fragmented and 6 tails, each from 30 to 44 degrees long. (Illustration from Milner's *Gallery of Nature*, 1860)

within 700,000 kilometers of disaster at the surface of the sun, and few manage to escape. All the fragments observed by the spacecraft in the past decade have disappeared into the cauldron after traveling from the depths of interstellar space, from far beyond the orbits of Neptune and Pluto, from about 30 billion kilometers away, seven times as far as Neptune, on journeys around the sun, which lasted hundreds of years and, in some cases, as much as a thousand.

The breakup of the most spectacular sun grazer in recent times occurred with comet 1882 II. It was that event which led Kreutz to examine the data and predict that its four fragments would return in 670, 770, 880, and 960 years' time. These four are likely to be annihilated when they next return. Yet not all of the sun grazers have fallen to their doom. In 1965, comet Ikeya-Seki lit up the heavens and was seen by millions of people in a rare, gregarious display from a member of the otherwise shy Kreutz family.

Ironically, the two spacecraft that discovered the majority (16 of 24) of this family of comets have themselves ceased to exist. The fully functional Navy satellite was used for successful SDI target practice, while SMM, as expected, returned to earth in December 1982. In NASA jargon, it "deorbited." Its burned and scarred fragments now lie at the bottom of the Indian Ocean.

Spectacular comet breakups have been seen quite frequently. Figure 4-4 shows a sketch made of the great comet of 1744 that suddenly broke into six distinct fragments.

TABLE 4–1 List of molecules found in comets

Two-atom molecules		
CH	CH+	C_2
CN	CN+	CO
CO+	CS	NH
N_2+	OH	OH+
S_2		
Three-atom molecules		
HCN	H_2S	NH_2
C_3	CO_2+	H_2O
H_2O		
Four-atom molecules		
H_2CO	NH_3	
Five-atom molecules		
CH_4		
Six-atom molecules		
CH_3CN	CH_3	OH

Note: From a list compiled by Lewis Snyder, University of Illinois. These molecules are mostly found in comet comas and are observed directly through their spectral signatures in the radio or infared bands.

The moral of this story is that when giant comets visit the sun they may break up into many large fragments, some of which can lurk about the solar system for tens of thousands of years. This may have happened around the time of the K/T event 65 million years ago and in chapter 8 we will consider the likelihood that something similar happened in recorded history. If a giant comet should make an unexpected visit in the next few years, and if it should break up, we can rest assured that our planet would be in for a very rough time, and the danger would persist for long into the future.

Despite their potential threat, without comets we would not be here today. Comet's contain a great deal of water and a lot of interesting molecules, listed in Table 4-1. In the days when the earth was still forming, comets deposited large quantities of both on the planet's surface; water for the oceans, and biologically important molecules with which to make living things.

Many of the organic molecules regarded as the precursors of life exist in comets and do not have to be manufactured on earth. After formation, the earth was bombarded by comets laced with many of the same molecular species found in dark interstellar clouds, which are veritable molecule factories. The full range of interstellar molecules, which includes over 60 carbon-bearing varieties known as organic molecules, is essential for life as we know it, and most if not all of them probably exist in comets, although astronomical technology has not yet allowed all of them to be recognized.

The point is that comets rather than asteroid bombardment of the early earth deposited the seeds of life as well as the water of the oceans on our planet. A dozen or so massive comets carry enough water and organic molecules to provide all the earth's water and biomass. Confirmation of the notion that biologically important molecules come from space came from the recent discovery that alien amino acids rained onto the earth around the time of the K/T impact 65 million years ago.

The chemistry of life is clearly related to the chemistry of space, and the mix required to make living things is common in the frozen depths between the stars, and apparently in comets as well. To discover just how similar the mixes are will require spacecraft to rendezvous with comets to sample directly the material of which they are made. That approach is to be favored over the alternative, which is to wait until after an impact to study what fell to earth.

5

THE BIRTH OF THE EARTH

*A*S we apprehend the likelihood of an almost inconceivable cosmic impact occurring again at some time in the future, it is worth considering how we got to be here in the first place. The quest for an explanation of our origins is, of course, as old as the ability of humans to conceptualize questions and consider answers. Our species has probably been able to do that for hundreds of thousands of years, since well before evidence of its ability to comprehend was etched in cave paintings, perhaps back in an age when stone tools began to be patiently chipped out of flint rock. But when questions about origins were first hesitatingly formulated, answers could only be invented. There was no way any human beings could have known back then what we know now about the nature of the universe and its contents.

Our collective ability to understand the world in which we live received an enormous impetus starting about 400 years ago when the scientific method for approaching reality was first practiced. That was when it was discovered that through experiment and observation, and above all through measurement, it became possible to unravel the secrets of the universe. That was when Galileo first pointed a telescope at the heavens, William Gilbert experimented with natural magnets, and Johannes Kepler discovered the laws of planetary motion. Since then, our species has gathered a stunning new perspective on the nature of this universe and its origins, a perspective that has relegated to the back burner of human thought most of the fantasies that have so long held sway over the human mind.

As a result of the high technology that has emerged during this century, scientists have learned to probe into the depths of matter and into the farthest reaches of space. In the course of this exploration, astronomers, in particular, have learned that the universe has its roots in awesome violence and that the birth of the earth and moon were accompanied by what, from our perspective, would be considered catastrophic events. Were anything remotely similar to occur today, all life on earth would be instantly terminated.

In popular imagination much of the cosmic violence seems incredibly remote, and it is. Stellar explosions that tear stars to pieces are seen on an almost daily basis, most of them in galaxies very far away. However, not for 400 years has anyone actually seen a star in the Milky Way galaxy die in a supernova explosion. We tend to regard such events as intellectual curiosities, even if we realize that without supernova explosions the stuff of planets and of life would never have been created at all.

To observe personally evidence for the violence that accompanied the birth of the solar system, all you need is a small telescope or a good pair of binoculars to look at the moon. Reflect on what you see marking its surface, craters on all scales (Figure 5-1), each crater created by an impact with a meteoroid, a comet, or an asteroid, most of them 4 billion or more years ago. But where did the comets and asteroids that scarred the moon originate?

Over twelve billion years ago, before there was a planet like earth, before stars took shape, there was an awesome expanse of nothingness, inconceivable vistas of emptiness beyond human comprehension—no energy, no mass, no time, and no space. Then a possibility emerged. A universe sprang into existence in dimensions of eternity that conscious inhabitants of planet earth would someday describe as space and time.

This is not the place to delve into the mysteries of cosmology, a subject I regard not only as highly esoteric but bordering on theology, although I accept that the best astronomical data to hand today imply that we live in a universe that had a very specific beginning, and which is now expanding, creating space as it goes. This is inferred from the observational fact that more distant galaxies are moving away from us faster than those relatively close by. These observations allow astronomers to figure out how long ago the galaxies would all have been in one place—that is, when the universal expansion began, although at the beginning of the universe galaxies did not yet exist. It appears that the expansion began about 12 billion years ago, give or take a few billion years, in that first moment in time called the big bang.

A few hundred thousand years after the beginning of the universe, it had expanded so much that the seething energy that marked its birth had cooled to the point where particles of matter were formed. That matter consisted of hydro-

Figure 5-1 The Eastern Hemisphere of the moon (on the sky, east is to the left) showing countless impact craters. Even the large dark areas, the so-called seas on the moon, or maria, are ancient impact structures filled by lava flows triggered by the impacts. Some of the prominent craters show distinct splash marks (streaks) where ejecta fell back around the crater. (Courtesy Lick Observatory)

gen with some helium thrown in for good measure, the stuff out of which the first generations of stars would be built.

Over time, swarming masses of gas gathered together, coaxed by the allure of gravity that binds all things in an endless dance. By the time the universe was a billion years old, vast clumps of gas had coagulated and shrunk to take on the eerie shapes we call galaxies within which trillions of stars emerged, hatched inside denser clouds of gas that were pulled together by the same primal attraction that nursed the galaxies into existence.

Within a shrinking gas cloud that was to form a star, a new source of energy, nuclear fusion, heated the incipient star and balance was reached between gravity's inward pull and the disruptive force of the additional heat from within. That balance defined the nature of the newly born star, which settled down to consume hydrogen in the nuclear furnace at its core.

Many galaxies' worth of stars formed when the universe was a few billion years old, and the light from countless stars streamed into space to travel for billions of years until it fell on telescopes on earth, telescopes built atop mountains or launched into space to see more clearly what clues the light beams carry, light guided by mirrors and lenses to fall on sensitive detectors and photographic film in the hope that the messages contained in those beams could be read and interpreted so as to tell us what the universe was like when stars and galaxies were still young. (Because light travels at a finite speed, looking into the depths of space at very distant objects means that we are really looking back in time as well.)

The young universe was filled with dense clusters of stars, born as families, clusters linked in liaisons that defined the galaxies we know today. Galaxies themselves swarm in clusters, each a vast hierarchy of structures whose nature puzzles the terrestrial astronomers as they try to understand what happened. Occasionally galaxies in clusters collide with one another to produce explosive consequences of a scale our imaginations cannot comprehend.

The early stars contained mostly hydrogen and helium, but within a star's fiery core transmutation of the elements occurred so that heavier elements were systematically cooked up according to the laws of physics, an orderly process that turned a star's core into a soup of the subsequent stuff of life. In this way, carbon, oxygen, nitrogen, sodium, calcium and all the elements lighter than iron were brewed in stellar ovens.

Fortunately for us, the most massive stars, the ones that cooked their contents fastest, because there was so much gas available in the furnace, ended their lives in dramatic explosions we call supernovae. In the initial moments of such explosive cataclysms, which send the remains of the star wafting into the depths of space, even heavier elements, such as iron, gold, silver, and uranium, were formed. The processed star stuff was then mixed with ancient gases to enrich the primeval mixture, setting the scene for the birth of future generations of stars that emerged, enriched by the heavy elements from previous cycles of starbirth and stardeath. After a few billion years, interstellar matter contained the mix of heavy elements that we now find within a typical planet like the earth.

As time flowed and space expanded to define ever greater volumes of the universe, vast star clusters gathered to form great galaxies and within one of those, the one we call the Milky Way, the scene was set for life to emerge on at least one planet, the one we call home.

The increasing presence of the heavy elements in space could be considered as pollutants of the pristine hydrogen and helium mix created in the early moments of time. It is also the most fundamental process that would some-day determine whether a species might arise on earth to contemplate this awe-some truth.

As the universe aged, more stars were born out of concentrations of the enriched mix of matter. All over the Galaxy something wondrous began to occur. Only the most massive stars died as supernovae. The vast majority of stars left the scene more gently. They ended their lives as red giants, swelling and cooling as the fuel for their nuclear furnaces ran out in the cores of those stars. These were the red giant stars. The largest among them, the supergiants, began to evaporate. Strong winds blew from their surface and these winds were enriched with ele-ments such as carbon and oxygen and nitrogen and silicon, which had been formed in their cores. These winds blew and created envelopes of matter that formed cocoons of gas and dust around the stars so that many of them disap-peared from sight. Only with the aid of infrared and radio telescopes have astronomers learned to pierce those cocoons to see what goes on inside.

On spatial scales larger than our solar system, a stellar envelope around a cool supergiant grows in density and is itself cool enough (a thousand degrees) so that the elements combine to form molecules. The envelopes that surround the aging supergiants became factories for hundreds of molecular species, a great many based on the carbon atom. These are the factories where the first steps in primitive organic chemistry, the chemistry of life, takes place.

Inevitably, the matter from those stellar envelopes, molecules, dust grains, and even comets—and, one suspects, lumps of solid, rocky or metallic objects, the detritus of a star's dying years—drift outward to infect all of space between the stars in the Galaxy.

The scene is thus set for the next phase: the birth of the sun out of a mole-cule-enriched cloud filled with dust and gas and probably cometary objects. When a star is born in this environment, it finds itself in a cocoon filled with a marvelous mix of material out of which planets can form. The material of the planetary system shrank under the pull of gravity and began to rotate as it did so. This phase is called a solar nebula.

The shrinking, rotating cloud took on a disklike shape and spun faster. At its heart most of the matter gathered and there a protostar emerged. Its temper-ature rose to a few hundred, then a thousand degrees, and it began to glow while around it smaller lumps of matter began to accrete within the gathering disk of material. Cometlike objects smash into each other and their matter coalesced to form ever larger objects. Asteroid-like objects, with perhaps more solid material in them than the water-laden comets, joined the wild dance as planetesimals, ever larger lumps of stuff, planets in their fetal phase, that gathered, collided, adhered, and grew ever larger.

Space around the evolving protostar became filled with countless objects the size of large and small moons. Still they collided, and still they collaborated to form ever larger objects, nebulous agglomerations of solid matter, water, molecules, and dust, growing inexorably into planet-sized masses that settled into orbit around the sun.

Each fledgling planet remained vulnerable to collisions with neighbors in close orbits. In a sense, the largest masses fought for dominance of their orbital zone about the star and in the process accreted the lesser masses that had, until then, lead an independent existence, growing at their own rate.

In such an evolving system of planets, around a medium-size star at the edge of the Milky Way galaxy, the earth was born. However, before it could be said to have been formed, a large neighbor, about the size of Mars, slammed headlong into the protoearth and turned it nearly upside down

The impact, the greatest ever felt by the planet, shook it from pole to pole and smashed it sideways so that after the chaos had subsided its axis was titled. This extraordinarily fortuitous circumstance would later favor the evolution of life, and in particular the emergence of a highly evolved species that 4.5 billion years later had migrated all over the planet. It managed to do so because of the favorable climate that held earth in its grip, because the planet was tilted in its orbit about the sun, and because it orbited not too close and not too far from that parent star.

Some of the debris from that awesome planet-shaking impact blew outward and was slowed by the planet's gravitational pull to be trapped in orbit to gather into what would someday be called the moon (Figure 5-1).

Far more important for life on earth, ever since then the moon has acted as a lock on the tilt of the earth's axis. Without the moon's steadying influence, the earth would wobble in vast excursions lasting millions or even hundreds of millions of years, which would drive the climate from one extreme to another. In that case, the climatic stability essential for the emergence and survival of life as we know it would have been nonexistent.

Back when the earth was blasted and titled on its axis in the process, sometime around 4.5 billion years ago, similar impacts between protoplanets and competing objects in nearby orbits occurred throughout the solar system. Uranus was impacted so violently that it ended up tilted sideways. Venus was struck so hard that it was began to spin in the opposite direction. Something smashed into Mercury with such violence that its outer layers were torn away and, lost to space, fell into the sun. And Mars? An impact tilted its axis but in the aftermath no moon was formed. Only two small objects, Phobos and Deimos (Figure 5-2), remained, but even they may have been captured since then. Mars was too small to hold enough debris to build a large moon to exert a controlling influence on the tilt of its axis. Today Mars is believed to wobble gently from side to side over periods of millions of years, driving the Martian climate through extremes that could not sustain life.

After the earth formed, the bombardment did not cease. Giant comets and

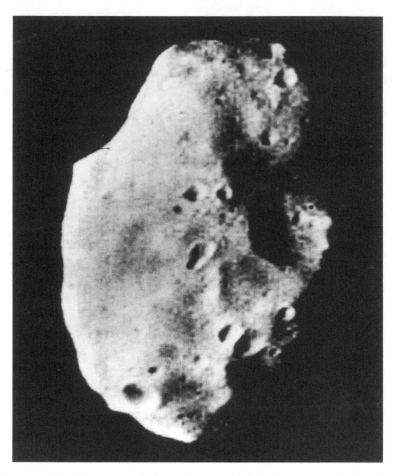

Figure 5-2 Martian moon Phobos photographed by *Mariner 9*. Phobos looks very much like the cratered asteroid Ida shown in Figure 3-5. At 20 by 25 kilometers across, it is about the size of comet Swift-Tuttle, which is due to pass close to earth in 2126. (Courtesy NASA)

asteroid-like objects continued to smash into the planet. Accretion, the process of planetary building, continued, as to a tiny extent it continues to this day.

The accretion by the early earth of comets through collisions played a key role in the history of life on earth. The comets brought the water of the oceans and the carbon-enriched material that would someday form the biomass of the planet, the stuff of life.

After the violence of planetary formation and accretion died away, simple living organisms emerged on the surface of the planet, in its oceans of cometary water. Those ancient organisms developed the power to reproduce and compete with each other for sustenance. Thus it came to pass that the scene was set for biological evolution to manifest itself. But all was not yet well.

James Kasting, geologist at Pennsylvania State University, and his colleagues have argued that an impact with a 100-kilometer diameter asteroid would have created so much dust as to shroud the planet for thousands of years, and so much heat that the oceans would have been boiled into the atmosphere. The water vapor would inevitably rain out until, many thousands of years later, the oceans were again filled with water. Nothing, except life in the deepest nooks and crannies of the primeval oceans, would have survived an impact of such magnitude. Following an impact with a large comet, the earth would have been sterilized. Yet such events were formative. They were part of the creative process that set the scene within which life would ultimately be able to establish a foothold in the face of continuing bombardment from space. We still cling to this uncertain perch.

At the time of writing it had just been announced that the evolution of the Martian atmosphere was also mediated by impact events. Mars is smaller than earth and because of its lower gravity, when impacts blasted gas and dust into space most of it never fell back. In the long run, the Martian atmosphere was eroded away into space.

Seen from this perspective, another question offers itself for scrutiny; what is in store for earth? Cosmic impacts will not become significantly less frequent in the next billion years. They may currently be less violent than they were 4 billion years ago, but they have not ceased. There are still a trillion comets in the Oort cloud, billions of objects in the Kuiper belt, and millions in the asteroid belt. All three are reservoirs for potential earth-crossing objects, and there are thousands of objects already in such orbits.

The earth and moon and all the planets continue to be smashed by objects from space. Rather intriguingly, moon rocks as well as fragments of Martian stones have been found among the thousands of meteorites found on the ice in Antarctica. That means that impacts on neighbors in space can create splashes large enough to eject material into an earth-crossing orbit where some of the fragments fall to the surface of our planet. This raises the question of whether terrestrial rocks could be splashed into space by an impact to later land on Mars, for example. That may not seem so terribly remarkable at first glance, but it is the added awareness that certain forms of terrestrial life can live in rocks deep beneath the earth's surface that challenges one's imagination. Certain bacteria can survive under extremely bizarre conditions deep in the earth's crust, such as a variety that has been found as much as 1700 feet beneath the surface where there is no oxygen or any obvious form of sustenance, except the minerals in the rocks themselves. Might such bacteria not be able to survive an equally strange journey through space until they land on another planet?

If we allow the possibility that primitive organisms can survive a 10-million-year journey to another planet, an age estimate based on the ages of certain meteorites, it is not so unbelievable to consider that they might survive an even longer journey to another planetary system associated with a nearby star. Or vice versa!

An image thus begins to form of life being splashed from one place in the

Galaxy to another, splashes triggered by collision events. The mind's eye then begins to take in the greater panorama of all stars and planets linked by comets carrying the stuff of life. This image was given its first hint of scientific backing when in May 1995, it was reported that bacteria that had lain dormant in a bee fossilized in amber for the past 25 million years had been brought back to life. Bacteria may be capable of surviving awesomely long journeys through the Galaxy.

From this perspective we might argue that the entire Galaxy is alive with primitive organisms that are ultimately all related, even if those relationships take a long time to become established. That does not mean that there are countless planets with life like ours, or even that human beings exist anywhere else.

Even if intelligent beings exist out there, they, too, will have to confront the hazards of surviving cosmic impacts. After all, twin civilizations would, by definition, only emerge on planets in which the environment is similar to that of earth, which includes the fundamental role played by impact events to drive evolution. I doubt that many, if any, such places exist. If they do, what is the probability that more than a few will have had the good luck to spawn civilization and a technological society between impacts violent enough to send them back into a stone age? I think we have been lucky to get where we are, and that precisely the same degree of luck (as would be required to create a twin civilization, or even a near replica) is unlikely to be common in the Galaxy. That is why I also think we need to reassess what it is we will do in the future by way of avoiding extinction.

Countless planets in the Galaxy are likely to be inhabited by primitive organisms such as those that were common on earth during the first few billion years of its history. The equivalent of what happened after that time, the great diversification of terrestrial life that began 570 million years ago at the end of the Precambrian, will surely have followed a different course on every habitable planet in the Milky Way.

But just what course did evolution follow here? Did things change slowly, so that the present is key to the past, a philosophical point of view known as uniformitarianism? If so, then change was gradual and uniform, and those geologic processes that shaped the terrain of our planet are then still observable at the present time. Or were the changes catastrophic, occurring in awesome and short-lived bursts? This diametrically opposed theory, called catastrophism, had its origins in the thinking of Georges Cuvier, who offered ample evidence for sudden change in the geological record. Such catastrophic changes may not be part of everyone's awareness; human lifetimes are too short to experience directly the awesome forces that occasionally operate to shape terrestrial terrain. In this way, we may spend our summer vacation at a beautiful campsite by a California mountain stream and never live to see it devastated by an infrequent torrential downpour that changes the shape of the valley forever.

The possibility that massively energetic events could alter the face of the planet in seconds or minutes is now being confronted in the impact scenario. More than that, we must confront the fact that the energy that shapes the earth has a substantial input from outer space. Not all the energy that drives continental drift necessarily comes from inside the planet. The earth is part of a far larger, dynamic whole that exists in the context of comet and asteroid strikes that spur geological change and biological evolution. All life on this planet lives within this context. We are not isolated from events in space. The threat of comets and asteroids is now recognized for what it is; the manifestation of a violent universe continuing to exercise its random influence on planetary events.

And so it came to pass that life began to stir on planet earth and the awesome power of elegant evolution began to sort the structures that emerged. Evolution offered patterns for survival and avenues for extinction. All this came to pass within a universe where vast numbers of lifeless bodies still swarm in abundance, some laden with water-ice impregnated with the building blocks of life, others more solid, pitted with craters telling of a history of countess impacts, even as life emerged on planet earth where comets and asteroid-like bodies continue to rain down, although few and far between.

Thus the stage was set for the emergence of a species that would someday learn to question the world and find answers. Four and a half billion years after the earth was formed, that species, *Homo sapiens*, discovered how to construct telescopes, devices for extending its senses into the depths of space. After a few hundred years of looking, it noticed that there was something happening nearby that required urgent attention. It discovered that the planet remains vulnerable to the same creative phenomena that set the scene for the emergence of life.

If there were other intelligent species in our Galaxy with the talent to observe what happens on earth, they might ask: "Now that human beings have begun to appreciate the dangers of existence in the cosmos, what are they doing to assure their long-term survival?"

Sixty-five million years ago a comet built of the stuff that spilled out by a thousand stars, a million shells of gas, a comet that may have traveled through eons of time, was carried into a trajectory that brought it into the solar system to slam headlong into the third planet from the sun. On impact, it burst asunder to spill its energy into the earth and its atmosphere, tearing the air outward in a vast plume of gas and dust laced with star stuff and mixed with molten rock and water from the planet's surface in a blast of violent noise and heat, a deafening crescendo of such magnitude that the debris encircled the planet with a blast wave that laid waste to virtually all that existed, flattening trees and forests and grasslands and creatures large and small. This input of extraterrestrial energy altered the chemistry of the atmosphere itself.

It took years before the creatures that survived, some of which lived below ground or in remote corners where their food supply was not destroyed, to emerge from the ruins. They roamed the devastated landscape wracked by flame and flood and found niches in which to flourish, oblivious of what had gone before. Those species pieced together a new existence, and they spread their influence as they diversified. Sixty-five million years later, a product of that speciation in the aftermath of the cataclysm learned to peer into the depths of space, and into the depths of earth, to recognize what had happened. In particular, it peered into the fossil record and learned to recall the past.

Even as that extraordinary species, *Homo sapiens*, figured out what was what, and how it came to be here, another comet was weaving its way through space. It settled into an orbit that brought it into a head-long rendezvous with the planet Jupiter in July 1994. Then the telescopic species on earth witnessed a momentous event (chapter 14).

Now we wonder if there might be another comet or asteroid roaming space, something left over from the formation of the solar system perhaps, on a course that may yet, in the not-too-distant future, rain sudden death from space upon the earth, and drive any of us lucky enough to survive back into the caves.

6

THE
NINETEENTH-CENTURY
PERSPECTIVE

*I*F comets and asteroids have a habit of
wandering dangerously close to the
earth, why wasn't the danger recognized a
long time ago? It was. In fact, before the beginning of the twentieth century the
threat of comets was taken for granted (asteroids had not yet entered the picture). Most astronomers in the nineteenth century accepted that the danger of
collision was so obvious that it hardly warranted argument. How they elaborated
on the danger varied from the understated, as in the case of Sir John Herschel
who in 1835 said that the experience of passing through a comet's tail might not
be "unattended by danger," to the dramatic, as we shall see.

In 1840, Thomas Dick, a well-known popularizer of astronomy, wrote a
wonderful book entitled *The Sidereal Heavens*. In it he reviewed all that was
known about the heavens, and did so from a theologian's perspective. This
meant that he repeatedly reminded his readers that the splendor of the night
skies was largely the responsibility of the "Divine." But then, if the existence of
planets, comets, nebulae, stars, the sun and moon could be attributed to God,
this raised a difficult issue for Dick. If comets were also part of God's plan, why
did the threat of impact exist? Surely God would never allow his creation to be
destroyed.

Dick did not shy from his predicament and began to search for an answer by conceding that little was known about the nature and origin of comets. At the time it was thought that the head of a comet probably consisted of "something analogous to globular masses of vapor, slightly condensed towards the center, and shining either by inherent light or by the reflected rays of the sun." The reason he could not be sure as to why the head glowed was because the means to study the properties of light had not yet been invented. That required the development of the spectroscope decades later, a device that breaks light into its various colors, which, when examined closely, can reveal the chemical signature of the object from which the light arrives. When astronomers looked at the light from comets, they recognized the signature of sunlight, which meant that comet heads glowed by reflected sunlight, and not of their own accord.

Even in Dick's time, comets were observed to manifest a great deal of structure, and this suggested to him and his contemporaries that there might be layers of clouds involved, just as on earth distinct cloud layers can sometimes be seen overhead. As regards comet nuclei, they were believed to be solid, although some were suspected of being transparent because of claims that stars had been seen through them. Comet tails, on the other hand, were regarded as most mysterious because of the way they pointed away from the sun and sometimes toward it.

About comet tails, Dick wrote with refreshing candor, "Of their origin or the substances of which they are composed, we are entirely ignorant, and it would be wasting time to enter into any speculation on this subject, as nothing could be presented to the view of the reader but vague conjectures, gratuitous hypotheses, and unfounded theories."

Dick was less reticent when it came to discussing the possibility that a comet had ever struck the earth, to cause a "concussion" as he put it. He was aware that comets traveled every part of the solar system and therefore it had to be possible that one might occasionally strike the earth. After all, close encounters in 1773 and 1832 were well known to astronomers of the time. He concluded that it was not impossible that a comet had come in contact with our globe, and quoted François Arago, the French scientist who in the early nineteenth century had calculated the chances of being struck. Considering a comet with a nucleus one-fourth the diameter of the earth (excessively large), and whose perihelion (point of closest approach to the sun) was inside the earth's orbit, Arago reckoned the chance of collision to be 1 in 281 million.

When we are confronted with something terrible and inevitable, even if the odds are that low, we tend to seek solace in beliefs that might lessen the harsh implications of reality. This is what Dick did and, in so doing, may have set the tone for future generations. He reminded his audience that the "Wise and Almighty Ruler" would never let anything like a comet impact befall our world, "without his sovereign permission and appointment" so that we "may repose ourselves in perfect security that no catastrophe from the impulse of terrestrial agents shall ever take place but in unison with his will, and for the accomplishment of the plans of his universal providence."

It is not clear how this was supposed to lend succor to those who confronted the terrible implications of comet impact. Even if it were considered the will of God, that would not make the consequence any less disastrous for the victims. Because Dick believed in the good nature of God, however, he concluded that such a catastrophe would never occur. But because an impact was within the bounds of possibility, it should be taken as a sign "that this earth and all its inhabitants are dependent for their present existence and comforts on the will of an Almighty Agent." If this Agent decided to do so, wrote Dick, it was clearly within his power to "easily disarrange the structure of our globe, and reduce its inhabitants either to misery or to complete destruction; and that, too, without altering a single physical law which now operates throughout the universe."

In other words, while he did not think that God would cause such a catastrophe, the very fact that it was in principle possible for a comet to strike the earth was a reminder of our vulnerability and helplessness in the face of the awesome power of the Divine.

Dick backed up his arguments about the *unlikelihood* of comet impact by resorting to Scripture. He could find no indication in the Bible that this would happen, at least not for ages. Since many important predictions mentioned in Revelation in the Bible had not yet come to pass, cometary impact was not expected. He estimated that an impact would only occur after "the Jews shall be brought into the Christian church," after "the idols of the nations shall be abolished," and after "wars shall cease to the ends of the earth." (From our modern perspective, it seemed unlikely that any of these goals will ever be reached!) The danger would come, he added, only after "the kingdom of Messiah shall extend over all nations," after "righteousness and praise spring forth before all nations," and after "the moral order prevails and all parts of the earth are cultivated and inhabited." Surely this would be an unlikely time for the catastrophe to occur. Why would God visit devastation upon such an Eden?

Lest there be any doubt in the mind of his readers, Dick summarized the threat of comet impact quite categorically:

> The believer in Divine revelation, therefore, has the fullest assurance that, whatever directions comets may take in their motions towards the center of our system, none of them shall be permitted to impinge upon our globe, or to effect its destruction, for at least a thousand years to come, or till the above and other predictions be completely accomplished.

Oddly, his final estimate of a thousand years before a major impact is not all that far off, as we will see in chapter 13.

With continued, albeit unconscious, prescience, Dick even speculated as to whether extinction of species might be related to impact events. He knew about the evidence for the sudden demise of mammoths, elephants, and rhinoceroses in Siberia, although in his day the notion of species was ill-formed, and the concept of mass extinctions barely considered. He concluded that he could not be certain that a sudden climate change in Siberia could be related to cometary influence, but that it was well within the bounds of possibility.

He returned to the theme of the cometary influence on life by asserting that even if one were to admit that there was an influence, this would not mean that he gives "the least countenance to foolish superstitions, or to the absurdities of astrology, since all that I would be disposed to admit in the present case is a purely *physical* influence; an influence that may exist, although we have not yet been able to discriminate its specific effects." Here he was even more marvelously prescient because scientists have since learned to discriminate the specific effects of past comet and/or asteroid impacts that were lacking in his time.

Dick offered this conclusion: "On the whole, we have no direct or satisfactory proofs that comets have ever come into direct contact with our globe, or that they have produced any considerable derangements throughout the planetary system; and whatever specific influence they may produce on our earth and atmosphere must be deduced from future observations." Such observations have now been made, and the reaction of the many of scientists who looked closely at the evidence has bordered on shock. The threat of comets (and asteroids) is not only real but devastating in its implications for our continued survival.

Thomas Dick listed tempests, hurricanes, volcanic eruptions, cold or hot seasons, overflowing of rivers, fogs, dense clouds of flies or locusts, the plague, the dysentery, the cholera, and other disorders due to comet visits. Almost anything calamitous that was not understood was blamed on comets, which meant virtually anything one wanted to explain away. Such "wild" beliefs have made it easy for modern scientists to scoff and explain away any attribution of terrestrial events to comet influence as so much superstition.

Even in the absence of collisions with our planet, the issue of a comet's influence on earth has another side. History clearly relates that comets have repeatedly filled people with fear. For example, in 1456 a splendid comet frightened Pope Calixtus so badly that "he ordered public prayers to be offered up in every town, and the bells to be tolled at noon of each day, to warn the people to supplicate to the mercy of Heaven. He at the same time excommunicated both the comet and the Turks, whose arms had lately proved victorious against Christians." So wrote Thomas Dick, who pointed out that the established tradition of having churches in Catholic countries ring their bells at noon dates back to that time. But why did people always react with fear? Was it because they didn't understand the phenomenon, or was it that at earlier times comets had, in fact, produced nasty side effects such as epidemics?

The association of comets with disease was the subject of a scholarly summary in 1829, when T. Forster wrote about the atmospheric causes of epidemics.

He claimed that most unhealthy periods were times during which a comet was seen; for example, the black plague was blamed on the comet of 1665. Thomas Dick was a sufficiently critical student of science to notice that if that plague had been brought to London by a comet, a similar outbreak should have occurred in the other great cities of the world. That didn't happen. He was acting as a good scientist when he wrote, "We err egregiously, in this as well as in many other respects, when we infer, from two contemporaneous events, that the one is either the sign or the *cause* of the other."

The most obvious criticism of the theory that comets bring disease is that every year there is a comet close to the heart of the solar system, either approaching or receding from the sun, which leaves plenty of scope for associating terrestrial events with comet visits. The possibility that something untoward may happen should the earth actually pass through a comet tail is another story, however. For example, before and after the K/T impact, alien amino acids rained onto the earth, and even in the absence of a devastating collision with a comet fragment, a close encounter might create significant terrestrial consequences. For example, the dust from a comet's tail, were it to rain onto the planet, might reflect enough sunlight for an extended period of time to cause severe cooling, and hence climate change.

But let us return to the deeper, psychological aspect of cometary apparitions. It is true, in Thomas Dick's words, "that the announcement of a comet has generally been received with melancholy anticipations, and the effects attributed to its influence have uniformly been of a calamitous nature." Why is this? Why are comets never heralds of bounteous times? Dick ventures to point out that if they did indeed portend ill for our planet, then, given that nothing is done without the guidance of the Deity, it would mean that inhabitants of other planets in the solar system were, at the same time, about to experience bad times. To Dick this seemed quite unreasonable, and "scarcely consistent with the boundless benevolence of the Divine mind." That the other planets were inhabited was not questioned by him, and to expect that all of them would be simultaneously "punished" was therefore absurd.

Although not admitting the influence of comets on our daily affairs, Dick exhibited marvelous New Age insight when he wrote, "The universe is one great whole, and all its parts, however remote, must be supposed to have a certain relationship with one another; and they may produce an influence, however small and imperceptible, on each other at the greatest distances."

In the background of the early study of comets there always lurked the awareness that the direct physical threat of impact was real. Simon Laplace the French philosopher and physicist knew of the danger and in the late eighteenth century went so far as to warn against pronouncing prematurely on impending disaster. After one such warning of cometary catastrophe, he claimed, which had turned out to be without foundation, a number of people died of fright and many women miscarried. He suggested that when a warning was issued in future it be done with calmness and with advice on how to avoid the worst consequences, a theme we will attempt to adhere to in this book. This is a very real consideration for those who are even now working on ways to avoid future impact.

Alexander von Humboldt in 1844 pointed out that Biela's comet was the first whose orbit was known to cross that of earth. "This position, with reference to our planet, may therefore be productive of danger, if we can associate an idea of danger with so extraordinary a natural phenomenon, whose history presents no parallel, and the results of which we are consequently unable correctly to estimate." He added that another comet in earth-crossing orbit, that of 1770, might someday collide with comet Biela to provide earth's inhabitants with the rare spectacle of a collision between two heavenly bodies. In this manner he shifted the focus from the likelihood of impact to the safe realm of distant space, predicting at the same time that something like that would never be seen for several million years. How wrong he was! The spectacle of a collision between two objects in the solar system was seen a mere 150 years later when Jupiter suffered the humiliation of comet impact.

In 1860 the Reverend Thomas Milner wrote a massive summary of all that was known about the natural sciences. He was no less awestruck by the cometary phenomenon, which he said had excited the attention of mankind more than any other. "Undoubtedly their sudden appearance, rapid movements, and occasionally extraordinary aspect, were calculated to awaken terror in ages of ignorance and superstition, and to originate the wild conjectures that are on record respecting their character and office."

He went on to present a vivid picture of the comet of 1680 which remained visible for several months. "The train [tail] reached to the zenith when the nucleus had set below the horizon, coruscations attending the whole length of the luminosity, giving to the phenomenon the aspect of a wrathful messenger, and not that of a tranquil body pursuing a harmless course." That comet was apparently the focus of much speculation and there were those who claimed that when its orbit was followed back in time it must have appeared at the time of the Deluge, a notion to which Milner paid short shrift. "Such chimeras deserve no serious notice" he scoffed. Yet such chimeras continue to be dragged out for scrutiny by those who think there may be truth to the rumor (see chapter 8).

Milner admitted that astronomers knew very little about the nature of comets, a state of affairs that is marginally better today. But he did agree with the widely held view that they appeared to be wholly gaseous, which suggested to him that if earth collided with a comet it would pass right through unscathed Therefore he scoffed at the implications that a comet impact could be associated with the Deluge, or even the depression that formed the Caspian Sea, another popular idea at the time.

A few decades earlier, Laplace (as quoted by Milner) had also written about the implications of comet impact:

> It is easy to represent the effect of such a shock upon the earth; the axis and motion of rotation changed; the waters abandoning their ancient position to precipitate themselves toward the new equator; the greater part of men and animals drowned in a universal deluge, or destroyed by the violence of the shock given to the terrestrial globe; whole species annihilated; all the monuments of human industry reversed; such are the disasters which a shock of a comet would produce.

Laplace was on the right track, but the details would not be filled in for several centuries, and then only reluctantly, after scientists identified the signature of comet or asteroid impact in the 65-million-year-old K/T boundary layer.

He elaborated his description as he groped for a deeper truth:

> We see then why the ocean has abandoned the highest mountains, on which it has left incontestable marks of its former abode. We see why the animals and plants of the south have existed in the climates of the north, where their relics and impressions are still found. Lastly it explains the short period of the existence of the moral world, whose earliest monuments do not go much further back than three thousands years. The human race, reduced to a small number of individuals, must necessarily have lost the remembrances of all sciences and of every art; and when the progress of civilization has again created new wants, everything has to be done again, as if mankind has just been placed on earth.

Despite the fact that scientists of two hundred years ago felt free to consider the likelihood and consequences of cometary impact, their speculations went largely unheeded. Even when they pointed out that for thousands of years of recorded history, cometary visits were looked at with more fear than pleasure, the danger of potential impact was ignored. Perhaps that was in part due to inherent skepticism, which persists to this day, regarding the role of nature in affecting the destiny of our species. Deep down we tend to regard natural disasters as mere inconveniences, and refuse to look at the larger picture, which places our species in something less than the center of the universe

In contrast, the role of comets in affecting human destiny was almost taken for granted by ancient peoples, who blamed comets for all sorts of wondrous things. Thus the Romans regarded the comet of 44 B.C. as a celestial chariot carrying the soul of Caesar. In Thomas Milner's words: "Cometary bodies have been deemed to be the vehicles in which departed spirits are shipped by their guardian angels for the realms of Paradise; and on the other hand they have been viewed as the active agents of natural and moral evil upon the surface of the earth, and been formally consigned to ecclesiastics for excommunications and cursing." He also took seriously the implications of impact and tried to placate his readers: "As to the near approach of a comet producing any great terrestrial change, such as deflecting our globe from its orbit by attraction, and by scampering off with it as a satellite, we have plain warrant to treat the assumption as romance."

To further alleviate any concerns, he told of Lexel's comet, which was seen in 1770 and then never again. It was thought to have passed within six times of the moon's distance of earth and its passage did nothing to the earth in its orbit, did not even add a single second to the length of a year, or cause a ripple in the tides, yet it was estimated to have been thirteen times as large as the moon. That size estimate was hopelessly incorrect and must have referred to the coma rather than the nucleus. That comet did not affect the earth; instead, the earth altered its orbit.

What Milner failed to appreciate was that the crucial aspect of cometary impact is not the ability of the earth's solid mass to shrug off the effect of the blow, but the inability of the atmosphere to absorb the energy of collision. A cometary impact will cause only the merest flicker in the earth's orbit, but its effect on the earth's atmosphere is something else entirely. The atmosphere cannot absorb the energy without substantial changes to its physical and chemical makeup. That is where the danger to civilization of comet impact lurks.

Milner concluded that his readers should relax their concerns and ended on an optimistic note: "It is no slight advantage to the moderns, that they can gaze upon such objects without anticipating disaster, and regard them as controlled by those laws to which their own world is obedient." This is the point we appreciate more fully, one that Thomas Dick also confronted. Comets behave according to nature's laws, which means that collisions with earth are not only a distinct possibility, but a probability (see chapters 12 and 13).

In 1860, O. M. Mitchell, the Civil War general who was a keen amateur astronomer, wrote, "Even in modern times, no eye can look upon the fiery train spread out for millions of miles athwart the sky, and watch the eccentric motions of these anomalous objects, without a feeling of dread." This in contrast to the movement of planets, which inspired confidence. He added: "It is useless to speculate with reference to the probable consequences of a collision, which there is scarcely one chance in millions can ever occur." He felt there was no indication in the nature of the earth's orbit that it had ever been disturbed by a collision.

One of the most extensive early discussions of the threat of comet impact was given by Amédée Guillemin in 1877. He had no hesitation in discussing the mechanical and physical effects of a collision with a comet. He built on what William Whiston, the theologian and astronomer, had said at the end of the seventeenth century about the likely connection between a comet and the Deluge. "According to Whiston, the famous comet of 1680, after having, 4000 years ago, produced the universal deluge, is destined to accomplish the destruction of the world, and our globe will ultimately be set on fire by the same comet which had previously inundated it."

Guillemin noted that "the amount of injury which the proximity of a comet or its collision with the earth would be capable of producing is undoubted." He quoted Maupertuis, who in 1742 wrote that terrible results were inevitable. Change in the poles and the earth's axis was expected. There was even a fear that if some comets came close enough to the earth, the planet would itself become a comet, which would cause the climate to wildly oscillate between extreme heat and cold in its new orbit, until such time as another passing comet restored it to its proper orbit. Maupertuis also thought that upon impact both the earth and the comet would be destroyed, but then would reassemble from the fragments. This may have happened back when the solar system formed, but it surely is totally unlikely now.

On a more realistic note, Maupertuis apparently realized that a smaller object upon impact would wipe out only a kingdom, which in some cases might be highly desirable, and, if the people were lucky, the debris of impact might even be replete with gold and diamonds to add to the joy of those who had gotten rid of a tyrant. Guillemin admitted that the evidence was not encouraging concerning the probability of finding gold and diamonds in comets, however, because the stones that had previously fallen from the sky were made of iron and nickel.

According to Guillemin, the astronomer Lambert had, in 1765, expressed his skepticism that comets could cause wars, famine, or the fall of empires, but he did appreciate that if one were to strike earth it would "occasion the most dire catastrophes" and even suggested that a winter of several centuries might follow. In view of the modern ideas of impact winter, his was a remarkably prescient insight. That instinct went astray when he speculated that collisions would actually be avoided by Jupiter's ability to deflect a comet from its course so as to avoid earth. This notion, too, is being seriously reconsidered by planetary scientists.

Guillemin did realize that "revolutions and catastrophes happen in the physical world" and reminded his readers that comet Biela had been shattered and that stars appeared and disappeared, such as the novae (new stars) of 1572, 1604, and 1866. (Those of 1572 and 1604 were in this century identified to have been supernovae.) Above all, collisions could occur because there was nothing in the known laws of physics that said they couldn't.

Guillemin had a chapter devoted to the consequences of impact based on the mechanical theory of heat, which lead many "philosophers" to consider that an impact would set fire to the globe "in which case we should perish by the shock and by fire." But this assumed the comet to have a fiery tail. If, instead, it were made of solid, cold matter then the energy of direct collision with a comet as massive as the earth was expected to stop the planet in its motion and the energy of impact would be transformed into heat. Hermann von Helmholtz, the German physicist, was one of those who calculated that, if this should happen, the earth would first fuse with the comet and much of it would be vaporized. Having been stopped in its path, the object that was formerly the earth would then fall into the sun.

Intriguingly, Guillemin went so far as to calculate the energy of impact, which, unlike our modern unit of megatons of TNT, was phrased in terms of the combustion of 5,600 globes of solid carbon each having a volume double that of earth. In some sense we have not come much further in 120 years. Our arguments about the likelihood and consequences of impact have resumed, this time set in a more realistic context, so we like to think, yet at some level we know as little as they did back then, although we at least know that comets are very much smaller than earth, which means that the planet as an inanimate object is safe.

The one topic that was touched upon by every writer during the past couple of centuries was the remarkable behavior of the comet of 1680. Whiston pounced on this comet as the cause of the Deluge. In his book, *A New Theory of Earth*, he tried to explain the geological upheavals recorded in Genesis. At first Whiston had invoked a comet to account for the Deluge but did not blame any specific wanderer from the depths of space. But the comet of 1680, whose period had been estimated to be 575 years, would have been around in 2344 and 2919 B.C. According to Hebrew texts, states Guillemin, the Deluge was supposed to have happened on November 28, 2349 or 2348 B.C.

Whiston's hypothesis was intriguingly seductive for his time. He imagined that the earth was itself once in a cometlike orbit, highly elliptical, which rendered its climate extreme and uninhabitable. When it was somehow moved into a more circular orbit there was a time when one hemisphere was faced toward the sun for the better part of a year and it was only after man had sinned that a small passing comet set the earth rotating.

It was further argued that God had foreseen that man would sin and from the start had placed a comet in an appropriate orbit so that it would arrive in time to do its damage. That occurred when the comet of a quarter of the earth's mass passed close to the planet and produced a prodigious tide, not only of the seas but of the solid ground which of course created widespread catastrophe. Thus "were all the fountains of the deep broken up" and when the earth passed through the comet tail its water and dust "opened all the cataracts of the heavens."

According to Whiston, the depths of the waters reached 10 kilometers. This would have wiped out all life on earth, and that is why Noah was required to account for the survivors. But at its next passage, this comet was supposed to cause the destruction of earth's inhabitants by fire because the encounter would then return the planet to an elliptical orbit, which would bring it close to the sun.

Guillemin was skeptical: "Such is the romance conceived by Whiston, a man of great erudition and science, but who shared the fault of his time in wishing to make his conceptions accord both with theology and astronomy." As we shall see later, that was not just a fault of his time (1696) but has bedeviled us ever since. Whether it is a fault is debatable, as we shall also see. Guillemin stated that Whiston's theory was untenable partly because comets were not massive enough to move the earth in its orbit, and, even if it were, the passage of the comet would be so rapid that it could not possibly affect the earth's orbit in the time available.

Speculation about the connection between the Deluge and comet impact was appropriate in Whiston's time, but only if done in the theologically correct manner. In 1694 Sir Edmond Halley also made the connection in a talk entitled "Some considerations about the Universal Deluge." This referred to the biblical flood which Noah doggedly survived in his ark. According to Whiston, Halley spoke with too much conviction because two weeks later he gave another talk, this one entitled "Some further thoughts on the same subject," in which he withdrew his earlier suggestion, an indication that he had come under pressure to recant. As we enter the twenty-first century and this issue again comes under scrutiny in light of modern findings about the nature and frequency of comet and asteroid impact, we should not be surprised if similar pressures begin to accompany the renewed interest in the cause of the Deluge.

Simon Newcomb in 1879 also wrote about the consequence of collision and steered far from the Deluge connection. He thought that passage through a comet tail would be harmless but collision with a solid metallic nucleus would be devastating.

> At the first contact in the upper regions of the atmosphere, the whole heavens would be illuminated with a resplendence beyond that of a thousand suns, the sky radiating a light which would blind every eye that beheld it, and a heat which would melt the hardest rocks.
>
> A few seconds of this, while the huge body was passing through the atmosphere, and the collision at the earth's surface would in an instant reduce everything there existing to a fiery vapor, and bury it miles deep in the solid earth.

Although not far from the modern simulations of an impact event, he drew a conclusion aimed to console his readers: "Happily, the chances of such a calamity are so minute that they need not cause the slightest uneasiness. There is hardly a possible form of death that is not a thousand times more probable than this." Newcomb thought that if one were to fire a gun at random into the air, the chances of bringing down a bird were greater than that of a comet striking the earth.

Camille Flammarion in 1894 also realized there was nothing in the laws of physics that forbade a collision. He did admit that collisions would produce unpredictable consequences since "these bodies are not...absolutely inoffensive." Also, a general poisoning of the human species, asphyxia, an unexpected explosion, or a mixing of some poison gases in the atmosphere could not be ruled out.

> But let us hasten to add that, notwithstanding the considerable number of comets and the variety of their irregular courses round the sun, it is probable that such a catastrophe will never happen till the natural death of the earth itself, because space

is immense, because our floating island moves with amazing rapidity, and because the point of the infinite which we occupy at each instant of its duration is imperceptible in immensity.

In other words, the likelihood of impact was just too small.

I find it fascinating that 150 years ago writers so freely speculated that comets might someday strike the earth. It was for them a logical and reasonable thing to do, in part because astronomers were thinking along those lines as well. They all struggled with the implications of the threat of comets and, like the theologian Thomas Dick, often wriggled out by resorting to beliefs about the nature of a Divine Being.

Around the time of Darwin in the middle of the nineteenth century, a conceptual shift caused scientists to reject the possibility that catastrophes (including comet impacts) could shape the course of terrestrial change, and hence of biological evolution. Catastrophism gave way to gradualism (also known as uniformitarianism), which subsequently exerted a stranglehold on scientific debate until the last quarter of the twentieth century. It also brought to an end for a period of time most speculation about the threat of comet collision until the phenomenon was finally placed on a scientific footing in 1980.

7

TWENTIETH-CENTURY RUMBLINGS

D URING the first century B.C., Lucretius wrote, "Legend tells of one occasion when fire got the upper hand. The victory of fire when the earth felt its withering blast, occurred when the galloping steeds that draw the chariot of the sun swept Phaeton from the true course right out of the zone of the ether and far over all lands." He knew about comets, which is why he said, "There is no lack of external bodies to rally out of infinite space and blast [the world] with a turbulent tornado or inflict some other mortal disaster." This awareness made him think that the world was newly made, and perhaps in some sense it is.

The wheel has come full circle. We now appreciate that the threat of comets and asteroids is real, although the distinction between comets and asteroids has grown blurred. What is no longer in doubt is that catastrophic impacts have occurred in the past, and that they will happen again. At the same time, the hypothesis that impacts and flood legends are related is beginning to experience a revival.

A chink in the dam of prejudice against the idea actually began to appear in the 1940's when two astronomers, Fletcher Watson and Ralph Baldwin, in separate books considered the implications of the discovery of near-earth asteroids (NEAs) and concluded that impacts were likely every million years or so. They were all but ignored.

In 1942 H. H. Nininger, the famous meteorite researcher, gave a talk to the Society for Research on Meteorites entitled "Cataclysm and Evolution." Because of his highly specialized forum, his remarks also went unheard in the wider astronomical community. He considered the danger following the close encounter with Hermes, the NEA discovered in October 1937 that passed within 670,000 kilometers of our planet, which can be compared with the moon's distance of 384,000 kilometers. (Oddly, Hermes has never been found again. Its rediscovery is one of the prizes that asteroid hunters strive for.) If, instead, it had "smacked the earth in a single lump," the consequences would "constitute a catastrophe of a magnitude never yet witnessed by man," said Nininger. He went further to suggest that asteroid impact might even account for the sudden disappearance of species. To him, lunar craters were a sign of the magnitude of past impacts, which meant that the earth, too, should have been struck repeatedly with consequences great enough to move continents. "Violent climatic changes would have resulted," he said in his talk, "locally at least, from the heat of the impacts and from changes in the content of the atmosphere," and "species would have disappeared and new ones would have developed to take their places."

Nininger's suggestion has turned out to be correct in many aspects, but back in 1942 there was no way to test his theory, and therefore it was treated as little more than idle speculation. That is the way of science. It is one thing to have a creative new idea, quite another to be able to demonstrate through experiment that it applies to the real world.

While the Velikovsky furor (see next chapter) was at its height, a remarkable little book was self-published in 1953 by Allan O. Kelly, a geochemist, and Frank Dachille, an amateur astronomer. Entitled *Target Earth*, it explored the possibility that impact events had shaped the planet and that the threat was still very real.

First, they noted that progress in the earth sciences has been bit by bit and that the evolution of theory had not been able to keep up with the flood of information then gathered by scientists. They suggested that this was because man was unable to view the earth as a whole, to see its larger features in relation to each other. "Perhaps, if he could view it from a distance, his theories of [the earth's] origin and growth might enlarge more swiftly." They added that all the theories about earth's evolution that had been taken seriously at any time involved collisions, but that few people ever considered that the threat was still present, largely because so few recent scars were known. This situation has changed dramatically in the 40 years since they wrote their book, but back then their suggestion that collisions continued to pose a hazard was ignored. Unfortunately, their suggestion was also confused by their wanting to explain too much with their model. They were aware of the criticism that they were trying to "take in too much territory" but insisted that there was evidence that the earth had repeatedly been struck by objects as large as several hundred kilometers in diameter. "Such cataclysms affected the whole surface of the earth and the life forms thereon."

So what did these authors suggest that was so outrageous that it caused their contemporaries to reject their ideas outright and force them to self-publish their book? This state of affairs did not make them happy: "The history of the science of geology has been one of continual strife and controversy, probably engendering more argument and bad feeling than any other scientific question." This was in part due to the clash between theology and science, as well as to human nature, which makes it difficult for us to give up a cherished thought. This was highlighted early in the nineteenth century when the opposing viewpoints encompassed by catastrophism and uniformitarianism were the focus of so much argument.

Kelly and Dachille thought that craters on the moon were the prime evidence for the role of impacts in shaping the surface bodies in the solar system. At the time that was a radical idea. They wrote their book when the origin of the lunar craters was still blamed on volcanism, but they argued that the data simply did not fit the facts. The lunar craters didn't look like volcanic structures.

They then tried to picture what would happen in a collision of a substantial object with earth. No matter whether it was on land or at sea, they expected a substantial, worldwide flood to occur. The orientation of the earth's polar axis might even change by a small amount, depending on where the object struck and at what angle. Today you will never hear an earth scientist suggest that a further tilt of the earth's axis is possible. It required the impact of a Mars-sized object to tilt the earth's axis in its formative years. The only effect an asteroid as large as a few hundred kilometers across will have on our planet will be to inject a great deal of energy into the atmosphere and to trigger global earthquakes.

Their insistence that the pole shift danger is real remains the weak point of their story. They thought that a collision with a very large asteroid would cause a readjustment of the earth's crust and violent volcanic action. An even larger impact would, according to them, scour everything clean down to bedrock and deposit the debris in low-lying basins. Vast floods would then smooth out this material. "The record in the rocks revealing these cosmic collision floods is world-wide, but the remains of the craters made by them are masquerading as mountains, ocean deeps, coast lines and other geographic features."

They were describing the very largest impacts, which modern discussions of mass-extinction events still do not take into account. Impact with a 10-kilometer object is bad enough, and no one is thinking about collision with a Chiron, 180 kilometers in size.

Kelly and Dachille consider the ocean basins as great impact scars. The curved coast lines are supposed to be the rims of the craters, but in 1953 when they wrote their book the theory of continental drift was not yet accepted. This accounts for the way continents around the Atlantic and Indian Oceans were once joined. The first ocean basin, the Pacific, may have been formed through impact about 4.2 billion years ago, at the same time that the large basins on the moon were made. Fortunately, a collision with an object large enough to gouge a hole ten thousand kilometers across has not occurred on earth in the past 4 billion years.

There are many other geological phenomena Kelly and Dachille blame on impact events, including the Carolina Bays in North Carolina. The origin of these circular structures continues to fascinate amateurs interested in impact craters but professional geologists say there is no evidence to support this hypothesis.

Kelly and Dachille were also ahead of their time in claiming that the extinction of species might be related to impacts. Mastodons were wiped out about 10,000 years ago when they seem to have been rapidly frozen. They suggested that that was a result of a shift in the earth's crust following an impact. The mastodon's distant relatives, Indian elephants, did not go extinct because they were too far from the new pole to be affected by the shift of the crust. With great prescience, the authors added, "To our mind it was not a case of the survival of the fittest but a survival of the lucky."

The situation with regard to frozen animals in the Siberian arctic is intriguing. They thought that the sudden freezing of the animals could only have come about as a result of the latitude change of the region where they lived. Also, the frozen creatures from formerly warmer climates are only found in one-half of the arctic hemisphere, which suggests how the pole shift happened and where the impact might have occurred. These animals are found where the ground is now permanently frozen a few feet deep, but when they fell, the ground could not have been frozen. Rather they fell and were then covered in water that froze and has never thawed since. In places the debris in which these animals are buried is more than 30 meters deep. "The recovery of perfectly preserved grass and flowers in the stomachs of some of the Siberian mammoth found frozen in the ice suggests that this great catastrophe took place in the summer time."

Cave paintings in France showed woolly mammoths, which were later found frozen in the Arctic ice 130 kilometers within the Arctic Circle. How else, but that the crust of the earth moved them there? I find this hypothesis unbelievable, even in view of what is known today about past impacts. After all, woolly mammoths were woolly because they already lived in a cold climate.

Target Earth quotes Thomas Crowell who described what was found in Alaska where countless specimens of creatures that once roamed a more temperate land were frozen. "It looks as though in the middle of some cataclysmic catastrophe of ten thousand years ago the whole Alaskan world of living animals and plants was suddenly frozen in mid-motion in a grim charade." When a bulldozer used to uncover gold veins dug into mammoth corpses, the smell of rotting meat spread for miles. That was the smell of 10,000-year-old mammoth meat exposed to air for the first time since the creatures died. If this hypothesis is correct, the ice layers in which the mastodons are embedded should contain traces of meteoritic material, unless of course the impact was due to an icy comet, in which case the deluge would have been even greater but with less solid material mixed in with the terrestrial and comet water.

Finally, the two authors approach the historical evidence, the sort of evidence that makes most scientists shudder, in particular as it pertains to the connection between a comet collision and the Deluge. "The literary treasures of so many peoples of the earth, treasures which were understandably preserved through pride of race and religion, are full of direct and indirect references to a collision-flood."

We continue to deny these legends as having any valid content, yet "With the passage of time the original messages in the legends, sagas, and writings that were passed down have been reduced to allegories or fairy tales by the indulgent superiority of the 'younger generations.'" I agree with the authors in asking why, if a universal deluge had not been a reality, this form for exterminating many people would have been chosen in the stories of so many cultures. Why not extra large volcanic eruptions or earthquakes, or extreme drought, flood, or blizzards, or whatever was familiar to a specific culture?

They made the very interesting point that, although the myths suggest that the flood was brought to punish humanity, efforts were made by survivors to save the animals. But the animals should not have been chosen for punishment, yet people such as Noah, are alleged to have saved them. A rain of 40 days and 40 nights, according to the Noah story, is roughly what might be expected from the aftereffects of an impact. The flood upon the ground is supposed to have lasted 150 days, which means that extensive rain was a crucial aspect of this flood. It was just more than a year when things returned to normal. That sounds a lot like the type of aftereffects now being discussed by the experts in modeling comet and asteroid collisions.

Target Earth ended with a remarkable suggestion that was 40 years ahead of its time:

> Therefore, it behooves us to consider ways and means to ward off a Day of Reckoning that may be set up in the mechanics of the Solar System and the Universe. In the increasingly numerous scientific discussions of man-made satellites or artificial moons, we see the beginnings of a system for protection of the earth. This system will require perpetual surveillance of a critical envelope of space with the charting of all objects that come close to a collision course with the earth. It will require, further, that on the discovery of a dangerous object moves be made to protect the earth. To this end might be used rocket "tug boats" sent out to deflect and guide the object from the collision course.

That was published four years before *Sputnik* was launched in 1957, at a time when few people other than science fiction writers dreamed of using artificial satellites for anything at all. Even more extraordinary is their foresight regarding the great flurry of events that would take place in the 1990s when a large body of scientists began to take the threat of comets and asteroids seriously: "We see then that the personal safety of civilized man extends outward from the police powers in his home town to a full and vigilant patrol of outer space."

This is now being seriously considered, as we shall see in chapter 17. The issue of whether the biblical flood can be related to comet collisions is also being discussed, but not yet within the "mainstream" of scientific opinion. That does not mean the hypothesis is without merit, however.

Kelly and Dachille's book made absolutely no impression on the scientific community. In defense of the "establishment," scientists could not do any more than speculate because as yet no hard data had been found to back up the hypothesis that the earth had been struck in the past. That meant that little had changed in a few centuries; all anyone could do was talk about the subject.

With the perspective of hindsight, further discussion of the threat of comets and asteroids could not be expected to move ahead without dramatic evidence to prove beyond doubt that our planet had been struck in the past, and that such impacts created demonstrable aftereffects. Evidence was provided by the discovery of excess iridium in the K/T boundary clays and highlighted by the further discovery of the impact scar beneath Chicxulub.

In 1957 M. W. de Laubenfels of Oregon State College wrote a brief article entitled "Dinosaur Extinction: One More Hypothesis," which turned out to be right on the mark. He wondered whether the dinosaur extinction might have been precipitated in the aftermath of a collision by an extralarge meteoritic object. He had harbored the idea since 1937, when the asteroid Hermes passed very close to earth. In addition, increased awareness of the damage done by the 1908 strike in Siberia (chapter 9) suggested to him that impact with a large object would produce devastating consequences, which he outlined with remarkable prescience, especially when considered in the light of recent research into the nature of comet or asteroid impact. His suggestion was phrased as an appeal to the paleontological community, in their main technical journal, and it is unlikely that many, if any, astronomers got to hear about it.

Another man ahead of his time, and hence also ignored, was the astronomer Ernst Öpik, who in 1958 published a peculiar little report entitled "On the Catastrophic Effects of Collisions with Celestial Bodies." In staccato style, he listed several points that were relevant if one was inclined to confront the possibility of such a catastrophe. He calculated how much rock would be melted or vaporized from impact of a body traveling at some velocity and with a certain diameter, to be chosen by the reader, and estimated the likelihood of impact with objects of a certain size. He concluded that collision with a 1-kilometer-diameter object might occur once every 29 million years. A half-kilometer object, now regarded as the size that would trigger the end of civilization, even if not large enough to cause extinction of species, was estimated to occur once every 590,000 years.

Öpik suggested that collision with a 34-kilometer object might be lethal on a global scale, wiping out all life. His prognostication agrees with more recent estimates of the consequence of being struck by such a large object. Öpik also went unheard.

Harold Urey of the University of California in La Jolla was very explicit in 1973 as to how he viewed the comet danger. He suggested that geological periods were terminated by collisions and regretted that he had first published his idea in the *Saturday Review of Literature*, not widely known as a place for scientists to share ideas. As regards the consequence of impact he said this:

> ...the energy was not dissipated in only vaporizing water or heating the atmosphere, or heating the ocean and so on, but the data indicate that a very great variation in climatic conditions covering the entire earth should occur and very violent physical effects should occur over a substantial fraction of the earth's surface. For example, the great seismic effects might initiate extensive lava flows.

Urey added that "the earthquake effect would be great in the immediate neighborhood of the collision site, and would be noticeable over the entire earth." He was also aware that survival would be a matter of good luck and concluded that "it does seem possible and even probable that a comet collision with the earth destroyed the dinosaurs and initiated the Tertiary division of geologic time," the K/T boundary.

Just three years before the discovery of the iridium enhancement in the K/T boundary layer became widely known in 1980, three geologists at Imperial College, London, John Norman, Neville Price, and Muo Chukwu-Ike, wrote a "speculative article" to suggest that large-scale cosmic impact features, which they called *astrons*, could be found on the earth's surface. These they attributed to meteoritic events that happened billions of years ago. They also realized that the earth should have received its fair share of impacts since Precambrian times and that some of the larger objects would have fallen in the water to produce "king-size" tsunamis possibly as high as 3 kilometers, which is precisely what has been found in recent computer simulations. They also realized that the earth's magnetic field might be affected, and that species would almost certainly go extinct as the result of such violent events.

Their final sentence bears quoting because it illustrates how tough it can be for scientists to grab onto a new idea that has not yet entered the mainstream of thought. "One is tempted to suggest that the demise of the dinosaurs could have been decided by a relatively small cosmic "catastrophe"—and even that mankind could suffer the same fate!" How right they were, as the world would soon learn.

For the first three-quarters of the twentieth century, the intrepid scientists mentioned here were lone voices in a wilderness where the subject of possible comet collision was not mentioned, at least not in "scientifically correct" circles.

Yet throughout the 19th century, the possibility of cometary impact had been a respectable subject for discourse, both by astronomers and popularizers of astronomy.

It continues to be difficult to accept that a comet or asteroid may yet slam into the earth in blind disregard of any life that exists here. We continue to struggle to acknowledge the threat of comets and asteroids. In view of what we will learn in the rest of this book, it is amazing how insightful Thomas Dick was in regard to a greater unity, a wholeness, to our universe, of which we are an integral part. That unity has been expressed in a manner that has led to the emergence of our species in the context of many catastrophic phenomena, not just cosmic impacts. Life on earth exists in a context that is as vast as the universe itself.

8

COMET IMPACTS
IN HISTORY

WHEN in 1980 Luis Alvarez and his team published their suggestion that the mass extinction event of 65 million years ago was triggered by the impact of a large asteroid, their notion was treated with scorn by many geologists, astronomers, and paleontologists. Since then, a growing consensus has developed among scientists from various disciplines concerning the relationship between the K/T event and impact. Following the identification of the associated crater in the Yucatan, a great deal of argument has subsided.

Even before the Alvarez study, two British astronomers, Victor Clube of Oxford University and William Napier of the Royal Observatory in Edinburgh, considered the possibility that there was evidence in ancient documents that suggested that lesser impacts had occurred in historical times. What they found was that "The dreaded expectation of fire from the heavens is a pregnant component of our intellectual heritage."

Their proposals were received with even less fervor than first greeted the Alvarez hypothesis. The treacherous quicksand between intellectual disciplines is where Clube and Napier have dared to roam, in this case between astronomy and history. For their troubles they have suffered the slings and arrows of outraged critics from both sides. Yet they have found some interesting patterns in history that bear repeating, stories that may tell us something about what the future holds in store for us.

The greatest barrier that stands in the way of Clube and Napier as they struggle to make their story heard may be the trauma suffered by astronomers in the United States in the 1950s. That was when the Immanuel Velikovsky affair stirred up painful controversy. Velikovsky was an amateur who dared suggest that there was evidence in ancient historical records, myth, and folklore for astronomical events that were not within the experience of modern astronomers. So far so good, and that is what Clube and Napier suggest as well. But Velikovsky added fanciful interpretations of the ancient myth and folklore while claiming his study was a scientific one. It didn't take critical readers long to recognize that Velikovsky was neither an able astronomer nor a historian, but somehow he managed to convince a publisher of his credibility. When his book *Worlds in Collision* was published in 1950, a great cry of horror and anger rose from the scientific community. The book-buying public, however, loved it.

Velikovsky claimed that certain biblical stories obviously told of great natural catastrophes, which is not so surprising to most people. But then he insisted that the Deluge was produced by a close encounter between earth and the planet Venus some few thousand years ago. Venus, in turn, was supposed to have been blasted out of Jupiter by a comet impact and, as it rushed past the earth to take up its new orbit, its tidal pull on the oceans created the Deluge. This nonsense was so absurd as to be laughable.

Velikovsky's book was filled with unsubstantiated wild ideas like this, which somehow caught the public's imagination. When the scientists reacted negatively it proceeded to sell in even larger numbers. The flames of public interest were further fanned by astronomers and historians who called for a boycott of all books by Velikovsky's publisher. Had they instead ignored his ravings the book might soon have been relegated to the trash pile. Instead, his sales and fame grew. Hordes of readers became dedicated followers of Velikovsky, who then wrote sequels to satiate their craving for more absurdities.

The reason scientists howled in anger was that Velikovsky had bypassed peer scrutiny of his ideas, which were marketed as if they were mainstream science. A fascinating account of both sides of the controversy is given in *Scientists Confront Velikovsky*. That book grew out of a public forum between Velikovsky, some of his followers, and a number of scientists including Carl Sagan. I happened to be in the audience to listen to the debate. Having been lured by Velikovsky's books as a teenager, and then later having seen how egregiously wrong he was, I was intrigued to see this ogre in person. He was clearly a powerful personality and a fanatic. He stated that his books were the only works of science that had required no revision in the decades since their publication, a claim that for me ranked as the greatest piece of self-delusion I had ever heard. The only books that require no revision are works of fiction. Anything that smacks remotely of science will require revision with passing years. It is the nature of science that we continually learn more about the subject of our studies, and therefore scientific texts are constantly in need of revision.

The point of mentioning the Velikovsky affair is that he gave this business of seeking clues to the possible occurrence of natural catastrophes in biblical stories, and in the myths, folklore, and legends of many cultures, a very bad name. Since then, anyone in the United States who in scientific company has dared hint at moving along the same path has been immediately branded as another Velikovsky, a label that is the kiss of death.

Victor Clube and William Napier were granted this sign of affection by most U.S. scientists when they dared suggest that there was abundant evidence in ancient history for catastrophes that can be associated with comet and asteroid impacts. Because tales of enormous floods and fire from the sky seem incredible to us, such stories have generally been interpreted as the amusing ravings of primitive people. We tend to interpret ancient legends, myth, and folklore in terms of the way the world appears to us today. It is only now beginning to dawn on us that the skies observed by the ancients may have been different from what we see today. This is the hypothesis that Clube and Napier explore so fruitfully in their two books, *The Cosmic Winter* and *The Cosmic Serpent.*

It is through the veil of trauma that surely surrounded the experience of ancient cataclysms that Clube and Napier have tried to look. They suggest that ancient tales about cataracts of fire from the sky may have been literal albeit primitive accounts of what happened. The stories may have been somewhat embellished with explanations involving a wrathful god pouring fire and brimstone onto the planet, or warrior gods clashing in the heavens spilling flaming debris onto the earth, an interpretation that would have been quite natural. But that should not cause us to loose sight of the underlying fact that something terrible was perceived. There were fiery things, such as fireballs, comets, asteroids, falling from the heavens in significant numbers, which caused widespread damage and took many, many lives.

Clube and Napier argue that, for good reasons, a preoccupation with the sky was an integral part of the earliest civilizations, and a fear of certain heavenly phenomena was built on an awareness that the sky presented a real threat to one's survival. "There was a clear perception that catastrophe might from time to time be visited on earth from above," they wrote. That perception is again with us.

An intriguing part of their argument is that in Babylonian times astrology dealt with omens, actual signs in the heavens, as opposed to horoscopes which are the bane of present-day, meaningless, astrology. The latter evolved out of omen astrology after the omens faded from the sky. In other words, the roots of astrology date back to an era when events in the heavens actually affected people. The threat of comets (and asteroids, which were not known back then) was understood to be real. People experienced impacts that shattered lives and even civilizations. Our distant ancestors apparently survived major disasters and cata-strophes of proportions that triggered the onset of dark ages during which soci-eties literally had to start from scratch before regaining some semblance of civi-lization.

According to Clube and Napier, there were times around 3,000 to 6,000 years ago when the earth may have suffered repeated impacts. Back then, ancient astrologers were gainfully employed doing what some planetary astronomers are again doing; searching for signs of impending disaster. Today asteroid hunters scan the skies to find NEAs (chapter 16), calculate orbits, and predict where the objects will be in the future. They also plan to sound warnings should rogue comets or asteroids threaten to wander too close to the earth. These astronomers are like the ancient omen astrologers who were engaged in trying to understand and forecast what they knew were very real threats from space. The technology used to facilitate the quest today just happens to be a whole lot better than any that existed thousands of years ago.

In Old Babylonian writing "flood and deluge were sent by the gods, along with equally catastrophic visitations of plague, drought and famine." This was back in the third millennium B.C. "The fact is that rulers and ruled alike in ancient Mesopotamia had absolute faith in astrologers" because the sky as understood by the Babylonians "harbored dangers which determined the des-tiny of nations." For them it was not a belief but a fact based on experience. Such a view of a threatening universe could only have emerged in the presence of real danger (Figure 8-1). It could not have emerged in an era when the heav-ens appeared as benign and unthreatening as they do today, "unless, of course, we are happy with the assumption that common sense and human courage are recent acquisitions."

Clube and Napier argue that the ancient obsession with astronomical events touching one's daily life could be understood in one of two ways: "Either the ancient sky behaved in some manner which *was* truly different from the present day, or modern man is indeed significantly more matter-of-fact and enlightened than his ancient counterpart."

It is not too surprising that the second alternative has been the one chosen by our contemporaries. The ancient view of a sky filled with arbitrary events capable of devastating civilization therefore gave way to one in which the uni-verse acts with unthreatening and clockwork regularity. Now we realize that this view, which can be traced to the time of Isaac Newton about 350 years ago, may be incorrect. The universe is anything but a simple mechanism. The Newtonian approach was seductive because it implied that we live in a highly predictable and hence comfortable world. Clearly the threat of random collision with a comet did not fit such a picture. To restore order, Newton imagined that comets, for whose existence he saw no good reason, were required to refuel the sun and they would do so without posing a threat to the planet.

Figure 8-1 A rock falling from the sky, a sixteenth-century woodcut that was meant to illustrate a biblical reference to the rocklike nature of Saint Peter. A meteorite this large, judging from the apparent perspective in this drawing, would do a great deal of damage to the planet. (From Camille Flammarion's *Popular Astronomy*, 1894)

The Newtonian world pictured everything to be under control, highly ordered, and unchanging, a stark contrast to what was believed in Babylonian times when the sky was filled with surprise and danger. The modern view gained strength from Darwin's concept of evolution, which invoked gradual biological change triggered by equally gradual changes in an otherwise benign environment. In the context of what has been learned about the role of impacts in triggering mass-extinction events, however, nothing could be further from the truth.

Clube and Napier stress that by placing snatches of ancient memory into a chronological sequence we may loose sight of the real story. We have become obsessed with the patterns *we* have stamped onto those stories and may have lost sight of the truths underlying the myths so that historians have tended to treat ancient myths as reflecting local idiosyncrasies. Ancient people, for some reason, are not trusted to have gotten it right.

When ancient people conjured up visions of fire from the sky, or of gods on chariots clashing in the heavens, we shrug our shoulders in amused tolerance. How primitive, how quaint. But what if they did see powerful forces racing across the sky and lashing at the planet. Perhaps they were trying their best to communicate about the terror they experienced when they saw fire falling from the sky. The task of astrologers was to predict the next, real danger. That may be why ancient Chinese astronomy focused so heavily on catastrophic events as well, on observations of "guest stars" (also known as novae or new stars, which actually included supernovae) and fireballs. These were often associated with catastrophic events on earth.

In ancient Mesopotamia, rulers and ruled alike "had absolute faith in astrologers and thus it was the advice of astrologers above all which was sought when it came to affairs of state." Today astrology has lost all meaning. The skies aren't filled with obvious danger any more. We live in relatively quiet times, but for how much longer? The danger that has now been identified is far more subtle. It is recognized in the traces of iridium found in the K/T clay and in the orbits of NEAs.

To put this more bluntly, can we be so sure that the Babylonian astrologers were not describing a dangerous sky? The Babylonians may have been afraid for good reason. When they reported cataracts of fire, perhaps that is what they saw—greatly heightened meteor activity that threatened terrestrial life. The weather was altered after such events, say Clube and Napier, who point to climate data obtained from observations of the movement of tree lines in the Arctic. Around 3000 B.C. and lasting for about two centuries, climate deteriorated on a global scale. Glaciers moved and there was major flooding in Mesopotamia and Egypt. There were changes in vegetation and ground cover, "including those brought on by extensive forest fires supposedly due to clearances undertaken in a period of presumed agricultural activity."

Clube and Napier do not try to prove beyond a shadow of doubt that cometary impacts account for many of the mysterious goings-on of ancient civilizations, but they do think that the possibility must be considered. This possibility has never been taken seriously by the scientific "establishment," largely because of a bias against the notion that celestial catastrophes played *any* role in human evolution.

To return to that period 5,000 years ago, Clube and Napier say: "Remarkably, the same epoch brings clear evidence of a surge in civilization; new skills, the appearance of writing and the establishment of a professional class, all coinciding essentially with the start of the historical era." They suggest that just as we now recognize the role of extraterrestrial intrusion into the machinations of nature, causing the extinction of species so as to set the scene for the emergence of new forms of life, so recent minor intrusions delineate changes in the evolution of human culture and civilization that are otherwise poorly understood. According to them, the clues are there to be seen, but only if we look at them with fresh minds, unhampered by our prejudices about the way primitive people tried to relate their most profound experiences.

> If the happenings [in the sky] were responsible in some way for assaults on the earth by cataracts of fire leading to a breakdown of law and order, we can expect some overall coherence in the pattern of world history: we can expect simultaneous collapse of empires, populations on the move at the same time, conflicts to be forced and new ideas to emerge, dominated perhaps by a deep fear of what the sky has in store.

They add that "this, strangely, *is* the pattern of world history and many a scholarly reputation has crumbled already in search of its underlying cause." Their own reputations have been sullied for suggesting that we consider that cosmic impacts are to blame, and mine may suffer for daring to suggest that they have a valid point. Yet their interpretation makes a great deal of sense. If a small comet were to strike in the near future, we, too, would see a breakdown of civilization and a gradual emergence of a new way of life as we struggled to start over again. We might also hope that as far as future generations are concerned, the description we then pass down to them regarding the cause of the chaos would be couched in terms that have a bearing on physical reality, rather than being metaphors of a poetic nature.

Clube and Napier offer examples of the disappearance of civilizations, such as the Minoan around 1450 B.C. and the Mycenean around 1200 B.C., which might be related to catastrophes caused by cosmic impacts. The aftermath of a comet collision would be expected to trigger a dark age that could last for centuries. Evidence for the cause of the decline of ancient civilizations is difficult to come by, but when contemporaneous pieces of the puzzle are compared, the scenario of impact disaster looks intriguing if not convincing. Often if not always there are references to astronomical events that tell of unwanted visitations, of fire in the sky, of wrathful gods, or of comets in whatever vivid form a particular culture assigned to such objects.

In later times, the nature of the sky gods began to change. "Subsequently, during the first millennium B.C., the malevolent gods appear to have departed, leaving a single god who became identified as both creator and director of a recurring universe whilst at the same time becoming himself increasingly remote and less tangible." This, Clube and Napier argue, occurred because the skies themselves cleared of the dangers that had seemed obvious to those who lived 5,000 years ago. In due course, the notion that such dangers ever existed was expunged from astrology, which was then left with a system of belief about the role of astronomical phenomena in guiding human destiny that made no sense any more. Astrology had to redefine itself, which lead to its inevitable decline into the useless set of beliefs peddled today in the guise of horoscopes.

I cannot recap all the arguments presented by Clube and Napier for visitations from the heavens, but mention must be made for circumstantial evidence that suggests that the earth may once have been in a different environment. This is related to the phenomenon of the zodiacal light, something very few of us have ever seen. It is clear from ancient texts and art that this light was often much more obvious than it is now.

The zodiacal light is produced by sunlight reflected off dust that lies in a plane around the sun defined by the planetary orbits and known as the ecliptic. That dust represents fine debris from long-lost comets and is studied by astronomers using space-borne telescopes sensitive to infrared (heat) radiation. Sometimes, when an active comet sheds a great deal of material, the zodiacal light can become very bright. It is currently so faint that it barely even receives mention in textbooks. Much of the zodiacal dust that is present now may be the leftovers from the breakup of a large comet around 3000 B.C. Ancient stories make repeated mention of the zodiacal light that was so prominent that it was confused with the Milky Way. "We are beginning to see, perhaps, hints of a night sky which was not the one we see now...," Clube and Napier wrote, and that "If the sky was in fact different in the first millennium B.C. and before, this might also account for the seemingly abnormal preoccupation of nearly all natural philosophers during the classical period with meteoric phenomena generally."

The consequences of impact were experienced as dramatically as those of earthquakes and volcanic eruptions. Since then, meteoroid impacts have decreased dramatically and now we find it difficult to believe that life was ever threatened, or that it might again be. "The spasmodic disintegration of a particularly impressive comet, then, could well have been responsible for a series of disasters in the Mesopotamian and Egyptian civilizations during the third millennium B.C., and then again for a subsequent series of disasters inflicted upon Minoan and Mycenean civilizations during the second millennium B.C."

Although such disasters are beyond our recent experience, their occurrence would account for the early obsession of ancient cultures with events in the heavens. But when these events dropped off in frequency, the new generations of philosophers increasingly viewed the ancient tales with skepticism, until today at least, when the possibility is once again forced into our consciousness. The discovery of earth-crossing objects and the signatures of serious impacts on our planet in the past raise a new specter: the threat of comets and asteroids to our civilization (Figure 8-2).

If comet or asteroid impacts occurred in historical times, there would have been occasions when such objects smashed into an ocean to produce tsunamis and flooding of surrounding coastlines. In that case, ancient cultures rich in folklore might have developed an oral history that contains stories of devastating floods, which they have. Virtually every culture has flood legends, the most famous one in the Western world being the biblical story of the Deluge that forced Noah to think about an ark.

Figure 8-2 Comet Morehouse photographed on September 30, 1908. Two exposures taken 3 hours apart show the tail separating from the head. A difference in exposure times produced the different length star trails since the telescope was made to track the comet. (Courtesy Yerkes Observatory, University of Chicago)

In 1992 Edith Kristan-Tollmann and Alexander Tollmann of the University of Vienna published a challenging article in an Austrian geological journal that took the Deluge-impact hypothesis a whole lot further. They drew together a wide range of information, both geological and historical, to conclude that about 9,500 years ago seven large fragments of a comet crashed in the seas around the world. (The number seven comes from Revelation 8 in the Bible, which is clearly a description of an impact catastrophe.) These impacts produced by the fragments of a comet triggered violent earthquakes and severe flooding, worldwide. Stories by eyewitnesses who survived those terrible times apparently correlate across cultures:

> The impact triggered global earthquakes of an unimaginable magnitude which deformed large landscapes totally, swallowed up islands in the sea, raised or broke down major mountain chains, moved the earth's crust like a storm-swept sea, crashed rocks, flung up the trees into the air, shot fountains of water into the sky—all of it reported by eye-witnesses.

Such a statement, unhappily for the Tollmanns, reminds American scientists too much of Velikovsky. This is unfortunate because the Tollmanns may have made an important discovery. They have analyzed legends and sagas from many cultures which suggest that impacts caused the ground to heave like a storm-swept sea; whirled trees, rocks, and people through the air; and triggered volcanism, specifically in the two Americas. A global wildfire was produced by flaming material that fell over wide areas. "Reports of those aspects come from the inhabitants of all continents, but chiefly from the Indians of America."

They point out that the comparison of events following the K/T impact is consistent with many flood legends. Myths and traditions about the Flood "have been preserved orally in the folk memories of most cultures around the world." Most of these stories have been collected only in the last few hundred years but their analysis has been more recent than that. Earth scientists have of course been very reluctant to attach any credence to the notion that such stories are anything more than imaginative tales unrelated to the experience of people in past ages. If, however, we allow for the possibility that they tell of events that were experienced by ancient people, a remarkably consistent story related to comet or asteroid impacts emerges.

A description of the possible existence of many comet fragments approaching the earth is found in Peruvian myth, which tells of a cosmic object that split into numerous fragments that looked like a group of stars. Other legends suggest that impacts occurred in the northern Atlantic, the western Indian Ocean, and the eastern Pacific off the coast of Central America. The impacts were in the oceans, although some debris has been found in Austria, which explains the interest of these authors in the phenomenon. A fragment fell in Köfels in the Ötz Valley in the Tyrol to produce extreme landslides.

The Revelation of Saint John speaks of the stars of heaven falling to earth as a fig tree drops its fruit (6:13) and similar tantalizing snippets lure one on, such as the claims by the Washo Indians in California, "the stars melted so that they rained like molten metal." Eyewitness accounts from various cultures suggest an impact in the northern Atlantic, and another in the Indian Ocean according to Indian tales, while an impact in the China Sea is apparently recorded in Chinese drawings and sculptures with the dragon symbol suggesting a comet.

According to the Gilgamesh epic, the Indian Ocean impact occurred around dawn in the Orient and other myths from the Northern Hemisphere suggest it was sometime in the fall. In the folk memory of the Chippewaya Indians in Canada, the great snowfall associated with a flood began in September, while Babylonian stories seem to confirm the approximate date. Southern Hemisphere stories would have the event occurring in the same time of the year.

No matter how tantalizing such tales may be, in the end there is nothing like some hard scientific data to suggest that something did occur around the time hinted at in mythology. The first clue comes from the age of tektites, which in Victoria, southern Australia, is about 10,000 years old. These were apparently produced by an impact southeast of Tasmania. Similar-age tektites in Vietnam appear to have been produced by an impact in the China Sea. In those areas of the world, myths tell of acidic, blood-colored rainfall. I must point out that the information about tektite ages is difficult to interpret. Based on radioactive techniques, they are 700,000 years old, and yet in Vietnam and in parts of Australia they are found at or near the surface of terrain consistent with having been deposited only 10,000 years ago. It may be that there are tektites of two distinct ages involved and, having read some of the studies of this tektite paradox, I would think that a lot more study is needed to resolve it.

The Tollmanns list other evidence that points to a date for the deluge around 7500 B.C., the most intriguing of which is the claim that a sliver of tektite was found in tree bark that clearly dates to 7500 B.C. Also, two Indian oceanographers, M. Shyam Prasad and P. S. Rao, found a tektite at the bottom of the Indian Ocean, which should have been buried beneath 1.4 meters of sediment if its age is 700,000 years. There are those who argue that tektites have moved up through the soil and sediments to confuse the issue, yet that alone should not cause us to dismiss the Tollmann's suggestions out of hand.

According to the Tollmanns, something dramatic happened around 7500 B.C. Based on what they claim to have found in the historical record, and comparing that with modern calculations of the consequence of a major collision, they conjure up the following picture of events. Immediately following a series of impacts, there was a great heat pulse produced by the fireballs. As a result of flood and fire, much of the world was left in ruins. Many of the ancient stories also tell of earthquakes of enormous magnitude. "The crust moved, according to these traditions, like the running high waves of the tempestuous sea, hurtled people on their faces, uprooted trees, crushed rocks, broke down mountain crests and raised them elsewhere, changed the landscape and submerged many islands in the Atlantic and in the Indian Ocean."

Wildfires are reported in tales from Indians in California, and similar stories from the Near East around the Euphrates. This then makes contact with what Clube and Napier have claimed. Also, aborigines in southern Australia still tell stories of terrible heat from the sky, consistent with the rain of "shooting stars" that has been described by scientists trying to model the consequences of impact.

After the fire came the floods triggered by tsunamis that lashed coasts, a result of the seven fragments falling into the oceans and seas around the world. This is inferred from the content of oral traditions that tell of water washing over a large fraction of the earth's surface. Only the highest mountain ranges and central and northern Asia were spared the floods. "According to some stories, the flood wave surmounted the Cordillera [western mountain range] of North America and penetrated deep into the continents."

Those parts of the continents spared from floods were deluged with torrential rains that brought mud and soot and acid, which in places turned the rain to a bloody color. "Countless myths report with horror that the evil wolf (Edda) or a pernicious demon devoured the sun, the moon and the stars." That was impact winter, which may have lasted three years according to the Tollmanns. In some places the torrential rains quickly turned to heavy snowfalls.

Terrible winds were also reported in various legends. There were tales of birth defects whose nature might be blamed on the noxious acidic rains, the expected increases in ultraviolet radiation following the removal of the ozone layer, and the suspension of heavy metals in the water following the rains.

Extinctions of species were associated with these ancient impacts. The mammoths appear to have been victims, but other species that died off around that time may have disappeared somewhat earlier, around 10,000 to 11,000 years ago, although the Tollmanns believe that a recent improvement in carbon 14 dating may cause those estimates to be revised downward to agree with the 7500 B.C. date of the floods. They also think that the ancient idea of the world passing through distinct periods, sometimes called "world years," can be attributed to the recurrence of cosmic catastrophes over periods of several thousand years.

The recovery of the planet following the flood impacts of 7500 B.C. is, according to them, described in Genesis. The sun and stars reappeared at the end of impact winter and the world was born anew. There would have followed a period of prolonged warmth, due to the release of greenhouse gases by the impact explosions. Life flourished, Eden-like. "This was the period when the roots of the great world religions were established with hell, purgatory, expiation and punishment, the fall from paradise into a world full of plague and trouble." Sacrifice, often human, became a way to mitigate against further impact catastrophes.

I am acutely and uncomfortably aware that to suggest that the Tollmann's have made a significant contribution to our understanding of the nature of recent impacts is dangerous, at least if I value my credibility as an impartial scientist reviewing all aspects of the nature of cosmic impacts. Several U.S. scientists who have taken it upon themselves to be spokespersons for impact events have expressed their distaste of how far the Tollmanns have pushed their interpretation of the data. However, none of them have read the Tollmann's full report, which so far has only been published in German and Dutch. (I have read the latter and think that there is more than a kernel of truth in their hypotheses.)

In chapter 12 I will argue that worldwide flooding from asteroid impact is expected every 5,000 years or so, and from this point of view evidence is expected in legends from ancient cultures for such events. Also, geological evidence of tsunami deposits should be sought. More intriguing is to ask how survivors would have dealt with such a catastrophe had it occurred 9,500 years ago. How might they have attempted to account for the resurgence of life after the flood? There is no way to unambiguously answer the question, although the point is that we may learn something if we dare examine in more detail the contents of ancient tales that seem to pertain to catastrophic events. Regrettably this challenge of interpreting ancient legends, sagas, and myths to prove a point is fraught with enormous pitfalls, as the sad Velikovsky affair illustrated. It is notoriously easy to read into ancient stories whatever one wishes to see. Yet we may also have been looking at those stories with a prejudice that prevented us from seeing.

So far there has been little support from the "experts" for the Tollmann scenario, but then they also reject what Clube and Napier have to suggest. The reaction to any proposal that we have anything of geological, paleontological, or astronomical value to learn from history is a direct parallel to the reaction met by the Alvarez team when they first dared to suggest a connection between astronomical events and mass extinctions. The prejudice of scientists against anything that docs not come from within their own discipline to account for a phenomenon that is believed to be within the domain of their discipline is considerable. Such prejudice reaches awesome proportions when it is suggested that fledgling civilizations may have suffered debilitating impact events as Clube and Napier have suggested, or that 9,500 years ago much of the world's population might have been wiped out, as the Tollmanns have claimed.

A great deal of research remains to be done to interpret ancient legends and geological data that may be relevant to the hypotheses offered by Clube and Napier as well as the Tollmanns. Such ideas will continue to be controversial but that, for me, is not the issue. We should allow the possibility that there is truth to the rumors and see where that leads. Perhaps impact events were more frequent than anyone has dared suggest in "establishment" circles.

In the final analysis, the message carried by the threat of comet and asteroid impact may not be whether we reach agreement on what happened in the past, but how we will deal with it in the future. By paying attention to the suggestions made by Clube and Napier, and the Tollmanns, our appreciation of the nature of the forces that shaped our existence may be enhanced and that, in the long run, will influence not only the survival of our species, but how we see ourselves in the greater cosmic context.

9

ON THE EDGE OF EXTINCTION

ON the morning of June 30, 1908, civilization may have suffered the worst piece of luck in its history. A small cometlike object exploded in the atmosphere above the Tunguska river valley in Siberia. It did little more than scorch and flatten trees for 20 kilometers in all directions and kill a thousand reindeer. However, if that object had struck a heavily populated region, we would not now dwell under any illusion concerning how close to the edge of extinction the human species actually hovers. Because the Tunguska missile missed a populated area, the threat of impact did not really begin to enter the public imagination until after the 1980 announcement of the discovery of the iridium in the K/T boundary layer.

Had the Tunguska object struck a large city, a million people or more might have perished, and the phenomenon would have raised everyone's awareness to the threat of comet impact. Instead, nearly a century later, the threat of comet and asteroid impact is regarded as little more than an interesting anecdote. Very slowly the nature of the threat is being recognized, but only because of the somewhat esoteric discovery that the dinosaurs were wiped out by a major impact 65 million years ago. Such huge collisions are infrequent, perhaps about once every 50 to 100 million years. It is the smaller impacts that pose the greatest danger, and they occur far more frequently.

About 800 years ago the South Island of New Zealand suffered widespread fires, which leveled the island and led to the extinction of the Moa bird. Maori legend says that a big explosion in the sky was the cause of the strange fire. Duncan Steel of the University of Adelaide and Peter Snow from Otago in New Zealand have pieced together a fascinating scenario that suggests that the Maori were correct. The fireball created by a comet impact may have ignited the forests of South Island.

Near the town of Tapanui in the province of Otago there exists a crater that geologists have been slow to identify as extraterrestrial in origin. According to Steel and Snow, it is not a conventional impact crater in that there is no evidence that a solid object struck the ground to create the indentation. Instead, an atmospheric explosion may have dug this hole, which is 600 by 900 meters and 130 meters deep. They add that, "There is...evidence of Maori myth, legend, poetry and song which speak of the falling of the skies, raging winds, upheaval of the earth, and mysterious devastating fires from space."

The connection seems to be confirmed by studies of fallen trees dating to that time. Within 40 to 80 kilometers of the crater, the trees point radially outward and just beyond that fallen trees point inward. The blast may have first flattened trees out to a certain distance and the firestorm ignited by impact fireball then blew down more distant trees as the flames burst back toward the crater area.

Place names in Maori tradition refer to events that are associated with the place, and the name Tapanui means "the big explosion" or "the big devastating blow." Steel and Snow add that a nearby region is called Otarehua, which refers to the Maori name for the star Antares in the constellation Scorpio. This suggests the time of impact was in the southern winter when the earth intersects the June branch of the Taurid Meteor Stream, a subject to which we will return to in chapter 11.

A similar story of a falling star bringing fire and havoc, killing many people, and depositing strange stones, apparently exists among the Australian aborigines of western New South Wales. Such tales offer interesting circumstantial evidence that the earth is under continued threat of bombardment from space, and that the planet is occasionally struck by objects that do extensive local damage.

The most famous case of an impact that caused great physical damage in an otherwise isolated region of earth was the 1908 impact at Tunguska. Although the explosion did not make a hole in the ground, it did blow down a lot of trees and kill reindeer. Every tree within an area about 60 kilometers across was flattened and scorched, but not burned. There, too, the trees fell in a pattern that allowed scientists to determine subsequently where ground zero was located.

Recent computer simulations account for the Tunguska event in terms of a 15-megaton blast created when a small comet or stray asteroid slammed into the atmosphere. A searing fireball ignited everything within miles and then the fire was almost immediately extinguished by the blast wave created by the explosion.

No sooner were the trees set ablaze than the fire was blown out and the trees toppled. It takes a very strong wind to blow down trees. Even major hurricanes do little more than uproot some trees. Tornadoes with winds of 320 kilometers per hour provide the best comparison. The point is that to flatten trees so that they fall neatly pointing away from the blast center, the winds would have been so strong that if a similar event were to happen over an urban area in the United States, say, every house would be collapsed in an instant with frightful loss of life. Estimates of the casualties that would result from a Tunguska-like event in a populated area in the late twentieth century suggest as many as 5 million dead (see chapter 12).

The full story of Tunguska has been slow in coming out. Two decades passed before the first scientific expedition led by the Russian scientist Leonid Kulik even reached the scene, although the fact that something dramatic happened in 1908 was almost immediately known all over the world. Seismographs sensed slight tremors, and barometric data recorded the passage of a pressure wave from the blast that rounded the planet several times.

Within a thousand kilometers of the impact, people saw a great flash in the sky. Dust from the comet's disintegration circled the globe and caused the sky in Europe to literally glow at night. In Scotland a few days later it was said to have been possible to read a newspaper outside at midnight.

In 1994 a report of what happened in 1908 was published in *Sky and Telescope*. The author, Ray Gallant, director of the University of Maine's Southworth Planetarium, had been invited by Tomsk University in 1992 to join in an expedition to the site. He summarized the fascinating results of patient work on the part of Russian scientists, who spent many years tracking down and interviewing survivors to find out what happened during and immediately after the impact.

A number of hunters and herdsmen beyond "ground zero" were knocked unconscious by the blast, and some of them were singed by the heat. Closer to the center of the explosion over a thousand reindeer were burned to a cinder. There were about a dozen teepee camps in the area at the time, but most of their inhabitants were out hunting well away from the blast area. Around the edge of the devastated area, people were blown off their feet and several were knocked unconscious. One old man is said to have died of shock, an understandable reaction considering what happened.

Quite unexpectedly, a huge explosion lit the sky and a fireball stretched nearly from horizon to horizon. The story is told that before another survivor had time to react, his clothes caught fire and then a terrible sound deafened him even as a powerful wind gust (the shock wave) blew him off his feet. Trees crashed to the ground as the wind blew out the fire, the flames unable to suck in enough oxygen to keep burning. The man was knocked unconscious and didn't appreciate his good luck that he had not burned to death or been blown away with the trees. When he recovered consciousness two days later, he was alone in the midst of a scene of utter devastation. Where once had stood his storage shed and his teepee, there were nothing but charred remains. That was enough to cause anyone's heart to reconsider its next few beats.

The 1908 object was probably of stony composition, which explains why it disintegrated high in the atmosphere. Ironically, the explosion of the Tunguska object may have occurred at an ideal height to minimize damage. Had it been made of more fragile matter it would have exploded higher in the atmosphere and the fireball would have ignited fires that would not have been blown out by a blast wave. Had it been made of solid rock or metal, it would have hit the ground and the 15-megaton blast would have made a crater the size of the Barringer and the explosion could have flattened a large city.

In 1994 two Italian scientists, Giuseppi Longo and Menotti Galli from the University of Bologna, and their colleagues claimed to have found evidence for excessive numbers of microscopic solid particles in tree resin from the Tunguska area that dates back to 1908. They used tree ring data from 1855 to 1980 and found the particles in the 1908 ring to contain iron, calcium, aluminum, silicon, gold titanium, and copper, all elements associated with metallic meteorites. They concluded that the impactor must have been an asteroid, unless a cometlike object can have an asteroid-like core.

An impact with the ferocity of a Tunguska is believed to occur once every 50 to 500 years, depending on whose estimates you accept. It is sobering to think how we would react to such an event in our modern television age, should it occur over a populated area. The fuss we make about natural catastrophes such as earthquakes or major hurricanes would pale in comparison. The damage beyond the immediate area of total destruction and carnage would itself be enormous. Days would go by before journalists and camera crews from nearby cities could travel into the devastated area, and then only by helicopter because flattened buildings would block all streets. A 2-kilometer-wide crater would be surrounded by terrain stripped of any buildings. Beyond that, damage would extend to hundreds of kilometers from the blast. The trauma felt by the survivors would outweigh anything experienced by those who have lived through devastating earthquakes. I think that we are quite unprepared for the horror that will be produced by even a small cosmic impact.

Four years after Tunguska, the *Titanic* sank with the loss of over 1500 lives. Its story is known to virtually every person in the developed world. Since then, great sums of money and a tremendous amount of human ingenuity have gone into making ocean travel safer. The same has happened for air travel. Airplane design stresses safety because a crash killing a few hundred causes great consternation. In contrast, the threat of impact is paid scant attention. The only significant impact event of this century happened in Tunguska and no one was actually killed. That is why the event may have been the worst piece of luck we have ever experienced as a civilization. Tunguska has become associated with irrelevance, yet a single impact like it today would cost as many lives as 20,000 airplane crashes or 3,000 *Titanics*. We have spent hardly any time or money (yet) on planning to avoid such a catastrophe or dealing with its aftermath.

The Tunguska missile struck the earth in 1908 and for every one like it there have been countless near misses. Only in the last few years has this been recognized.

On March 16, 1994, I received an e-mail message from Duncan Steel in Australia. I had been in contact with him regarding research for this book and he mentioned a close encounter with a near-earth asteroid (NEA) that occurred on the previous day. He assumed that I knew about it, but as far as I was aware the U.S. media took no notice. He forwarded a statement he had given to the Australian media.

> About six hours ago the earth had a near-record observed near-miss by an asteroid. The miss distance was about 180,000 kilometers, which is less than half the distance to the Moon. The object is only about 10–20 meters (30–60 feet) in size. Its name at this stage is 1994 ES1. It was discovered by the Spacewatch team (University of Arizona) at Kitt Peak National Observatory, near Tucson, Arizona.
>
> If it had hit the earth it would have done so at a speed of about 19 kilometers/s (or 44,000 mph). Unless it is made of solid nickel-iron (as are many meteorites) it would have exploded in the atmosphere at a height of 5–10 kilometers. The total energy released would be equivalent to a nuclear explosion of energy about 200 kilotons (around 20 times the Hiroshima bomb).

He appended a note saying "I had imagined that the U.S. media would be full of it by now....I have already done 3 radio interviews this morning (and I've only been at work 30 minutes)." Nothing of this close pass made the news where I lived at the time, Memphis, Tennessee. But beside the close encounter, there was another story that took shape, one that is known to few people other than those directly involved.

Whenever a new asteroid is discovered (Figure 9-1), the International Astronomical Union's Minor Planet Center at the Smithsonian Astrophysical Observatory in Cambridge, Massachusetts, issues an announcement. It usually gives positions where astronomers can point their telescopes in order to observe the new object, so that confirmation of the discovery can quickly be made and the orbit can be calculated. That, of course, is the first step in determining whether an impact is possible.

On this occasion, Minor Planet Electronic Circular 1994-E05 was sent to announce the discovery of the asteroid, which also appealed for immediate observations to confirm the orbit. Brian Marsden took the opportunity to use 1994 ES1 in an experiment to determine what problems might arise if an asteroid appeared to threaten earth and immediate follow-up observations were

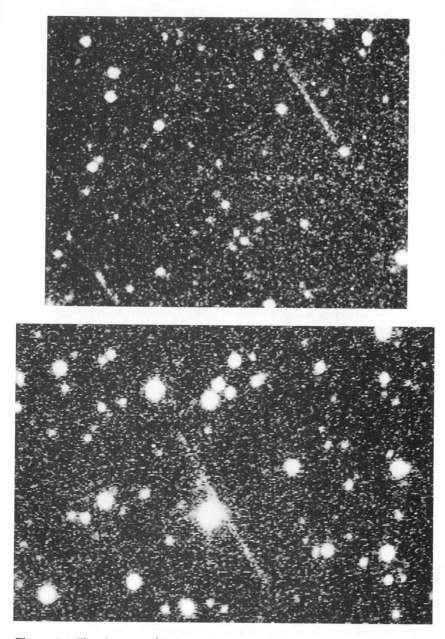

Figure 9-1 Two Spacewatch images showing newly discovered asteroids moving across the background star fields. The asteroids leave streaks indicating how far they have moved during the data taking. That allows a first estimate of their approximate distance to be made, assuming that they are traveling like well-behaved asteroids. The image below with the single long streak shows asteroid 1993 KA2. The other frame shows 1994 GK (long streak) and 1994 GL discovered together. The different streak lengths indicate that the objects were at very different distances, even if they were all in the asteroid belt. (Courtesy Spacewatch, Lunar and Planetary Laboratory, University of Arizona)

required. Since the detection of this NEA was made at Tucson, Arizona, during routine observations, follow-up had to be done somewhere else as night moved around the planet—for example in Japan, New Zealand, or Australia. So Marsden sent an e-mail message to four key observatories he knew would be able to obtain the required data. The results of the test were reported in another circular:

> Discovered 1994 Mar. 14 by Spacewatch.
>
> Observer D. Rabinowitz. Measured by J. V. Scotti.
>
> An initial orbit determination already on the discovery night showed that 1994 ES1 would pass only 0.001 AU [150,000 kilometers] from the earth a day and a half later (on the Ides of March), at which time it would be moving rapidly across the sky for observers in New Zealand, Japan and Australia, prior to running into daylight over the Indian Ocean. Four astronomers in those countries were therefore supplied with an initial ephemeris [orbit data] and alerted to the likelihood that it would be improved as the result of further Spacewatch observations on the night after discovery. *This exercise seemed to be a useful simulation of what could be done if the object were perceived to be a real danger to the earth.* Unfortunately, the exercise failed....Failure was due to communication delays with Japan, bad weather in New Zealand, and an observatory in Australia missed the object by 0.5 degrees because of inadequate allowance for parallax. The Minor Planet Center attempted to provide topocentric ephemerides to the Australian observatories, *but these could not be communicated in time—something that has now already been rectified.*

My use of italics highlights the drama. Technical problems delayed the communication on its way to Japan, bad weather in New Zealand prevented any observations, and an error in pointing the telescope at one of the Australian observatories meant they observed the wrong area of sky and didn't see the approaching asteroid at all. The fourth addressee of the original electronic announcement never received the warning because students had tied up the relevant computers for 10 crucial hours. Were they really just working hard?

The last phrase in the union's circular quoted here, concerning the fate of the urgent message that "could not be communicated in time—something that has now already been rectified," says a lot. The test was a simulation of what might happen if an NEA were seen to be heading for earth. How fast would astronomers react to produce an accurate prediction of where the impact would occur? After all, the fate of millions might yet hinge on how such alerts are handled in future. The test led to the creation of more secure computer links. Of course, if it had been a real alarm the message would also have been phoned through to the other observatories to make absolutely sure that it got through. The experiment highlighted teething troubles to be expected once a defense system against NEAs is permanently put into place.

Just how common are close encounters with near-earth objects such as the one that hurtled by in March 1994? It turns out to be much more frequent that one might imagine. Every few hours an object about 6 meters in size passes *inside* the lunar orbit, and every year one such object hits the earth to produce a 20-kiloton explosion. With the discovery of the new asteroid belt near the earth, the frequency of impacts by these smaller objects should perhaps be once a month, according to David Rabinowitz of the Spacewatch team. Why then, don't we notice more impacts? Based on the available statistics, astronomers have wondered about this for some time, and in October 1993 a remarkable report on impacts was declassified by the U.S. Department of Defense.

For several decades infrared satellites deployed in geosynchronous orbits and designed to search for the telltale heat signatures from enemy rocket exhausts have picked up lots of bright heat flashes produced by objects smashing into the atmosphere. Over a period of 18 years, 136 such events were detected, which represented only 10 percent of the total estimated to have occurred during that time. That averages to about 6 per month that were not seen from the ground because their energy was mostly released in the heat (infrared) band of the spectrum. The flashes lasted a second or so and unless someone happened to be looking in precisely the right direction at precisely the right time, there would be no chance of having seen anything. (The brightest flash seen from orbit in these 18 years was probably a 5-kiloton event.)

On August 3, 1963, there was a strong blast in the atmosphere between South Africa and the Antarctic that was detected by military satellites. It was equivalent to a half-megaton blast, which raised speculations that it was a small nuclear device tested by the South Africans in cahoots with the Israelis. It is now suspected to have been an asteroid or comet collision (unless you know something to the contrary).

Apart from the many near misses or strikes from small objects, there have been some interesting close calls by larger objects in recent years. Most of them are by asteroids that would do little more than create a problem if you were standing about where they wanted to make a small crater. For example, on December 12, 1994, we read in our local newspaper that "Asteroid almost hits us, expert says." The object was the size of a school bus and it was discovered 14 hours before it flew by within 105,000 kilometers of earth. That was very close indeed, "the astronomical equivalent of a near-collision of two cars in an intersection" it was claimed. Close enough to shout at the other driver, but not close enough to call the insurance company. Impact would not have been too dramatic, about a single "Hiroshima"(10 to 20 kilotons), which has become the unit for describing small-impact energies. This object would probably have exploded quite harmlessly high in the atmosphere.

On June 21, 1993, one U.S. newspaper reported "Asteroid's Close Pass Went Undetected." A 10-meter object (another school bus?) passed within 150,000 kilometers. It was discovered hours after it passed the earth on May 20.

Had it struck, it could have dug a nice crater, or caused a large wave. It was also discovered by the Spacewatch team.

On January 18, 1991, Spacewatch saw a similar-sized object pass within 170,000 kilometers of earth and its impact energy would have been about four to eight Hiroshimas. This was another member of the fleet of school-bus-sized objects that seem to make up the new asteroid belt referred to in chapter 3.

A larger object about 200 meters in size with a very peculiar orbit was discovered by Robert Jedicke of the Spacewatch team on February 3, 1995. Known as 1995 CR, it follows a highly eccentric path that crosses the orbits of Mercury, Venus, earth, and Mars. This type of orbit is highly unstable (chaotic) and before long, at an unpredictable time in the future, 1995 CR will smash into one of these four planets, or the sun, or will be thrown out of the solar system.

Most of the school-bus-sized NEAs are little pellets compared with the object that missed us by 650,000 kilometers on March 23, 1989. Had it struck we would not be around to reflect on anything, let alone close encounters. The object 1989 FC, or Asclepius as it is now called, passed just beyond the moon's orbit, which sounds like a nice, safe distance. But it missed by only 6 hours, which sounds very close. No one is quite sure how large it was (about 300 meters across) and the impact would have produced a blast of between 1,000 and 20,000 megatons, the uncertainty related to not knowing just what 1989 FC was made of. No matter what the details, it was firmly in a size category for objects capable of creating a global catastrophe, and it reminds us that we are perpetually teetering on the brink of extinction, whether we like to think about that or not.

The awareness of dangerous close encounters with NEAs would have grown more rapidly if the 1937 close pass of Hermes, an earth-crossing asteroid a little larger than 1989 FC, which passed at a distance of 650,000 kilometers, had been seen in the light of modern evidence for the relevance of past collisions.

The search for NEAs only began in earnest a decade or so ago, thanks to the personal dedication of Eleanor Helin and Eugene and Carolyn Shoemaker at Palomar Mountain, Tom Gehrels and his Spacewatch team in Tucson, and Steel and his colleagues in Australia. That is why the list of recent close encounters is sparse. No one has any idea how many other NEAs have passed only to continue on their way unseen. Statistical estimates can be made, but statistics are not data. What's more, 1989 FC was not seen coming; it was only seen going. No one happened to be searching in the right direction until after it had whizzed by our planet.

What about the future? Do we know of any comet headed our way? We do, and it is a very large one. The good news is that it will be a while before it comes back. The culprit, comet Swift-Tuttle has been described as the single most dangerous object known to humanity. It will remain so for 10,000 to 20,000 years, after which its orbit is likely to deteriorate so that it will either fall into the sun or be thrown out of the solar system, provided it doesn't hit earth before it does that. It is the parent comet of the Perseid meteor stream and was first seen in July 1862. Within a month it became a spectacular sight with a 30° tail that grew as bright as some of the brightest stars in the sky. The comet passed within 50 million miles of the earth, not close by NEA standards, but it was in an earth-crossing orbit and was supposed to return in 120 years. It didn't.

In 1862 the comet was observed from the Cape Observatory in South Africa and during the last week it was seen its position seemed to be off by about 10 seconds of angle, a tiny amount by everyday standards, but a lot for astronomers. The sudden shift in position became known as the "Cape effect." Those who since tried to predict the comet's return struggled to include the Cape data, because they could have a great effect on the orbit calculations.

The cause of the Cape effect is suspected to be related to comet Swift-Tuttle's proclivity to emit jets of gas that cause it to move sideways and thus to follow a slightly different path, just like comet Halley and the legendary Schwassmann-Wachmann. In the jargon of the space age, those little bursts produced midcourse corrections. Unfortunately, their effect is difficult to estimate and impossible to predict. On August 23, 1862, French astronomers saw Swift-Tuttle's tail undergo spectacular changes and material seemed to jet out of the comet's coma. That was one of the events that affected the orbit. This vagary makes the return of Swift-Tuttle difficult to predict. Given that it is an earth-crosser, its return is a matter of more than cursory interest.

The comet is unique in being locked into a 1:11 resonance with Jupiter, which means that for every one of its orbits around the sun, Jupiter completes 11 orbits. Computer calculations of Swift-Tuttle's orbit therefore have to take Jupiter into account at all times, and this is mathematically very tough.

On the basis of the 1862 data, Brian Marsden estimated that the comet should have returned in 1981–82, but it didn't show. Apparently the Cape effect had ruined the prediction. Comet orbits can be refined if the comet makes repeated visits to the sun, but finding which past comets could be associated with Swift-Tuttle turned out to be very difficult as well. If a comet does not return on a regular enough schedule, there is almost no way to link past apparitions to the object of study. Donald Yeomans and colleagues at the Jet Propulsion Laboratory in Pasadena managed to identify Swift-Tuttle sightings in 69 B.C. and A.D. 188, when it came particularly close to earth, but they could find no records of earlier close passes, such as those that should have occurred in 447 and 574 B.C. Those data were insufficient to predict its most recent return.

Comet Swift-Tuttle was finally rediscovered in September 1992. Its orbit had been affected by the jetting of material from what must be a rather fragile body. Soon after its reappearance it was recognized that it also made a close pass in 1737 under the guise of an object named after the Jesuit priest, Ignatius Kegler.

Yeomans and his colleagues have enough data to predict Swift-Tuttle's next pass. It will miss earth by 23 million kilometers on August 5, 2126, which translates into about 14 days with the current uncertainty on this estimate of about a day each way. When the first estimate was made, it appeared that the uncertainty allowed for a possible impact and the news was immediately broadcast around the world. "Life on earth threatened with extinction" blared the headlines. That now seems unlikely, although Brian Marsden is not yet convinced that the predictions are accurate enough for inhabitants of earth in 2126 to rest in peace (or will they?).

In early 1993, the comet ejected material that changed its path once again, albeit very slightly. But such slight changes amount to a great deal after 136 years. This time the Cape effect was observed by astronomers in New Zealand. The role of these deviations is unknown and therefore astronomers continued to monitor comet Swift-Tuttle as it recedes into the distance among the stars in the Vela constellation. Disturbances to its orbit are less likely farther from the sun, unless collisions with meteoroids fracture their surfaces to allow volatile material to escape and steer the comet in a new direction. Position data will be gathered until 1998 when the comet will become too faint to be seen, even with the world's largest telescopes. Then the very best predictions for its return will be possible.

Comet Swift-Tuttle is really a very dangerous object. It is about 24 kilometers across, far larger than the dinosaur killer, and upon crossing the earth's orbit in A.D. 2126 it will be traveling at nearly 61 kilometers per second. The energy of impact would be in the range 3 to 6 billion megatons which compares with a mere 100 million megatons of the K/T comet, which led to the eventual extinction of about 50 to 60 percent of species on earth.

This number is so large that we cannot begin to fathom its magnitude. Suffice it to say that based upon what is known about impact physics and the extinction of species, if comet Swift-Tuttle were to strike the earth, very little would be left alive. Life might even have to start all over again, stimulated by bacteria, viruses, and cockroaches, all of which seem to have a dramatic talent for living in outrageously hostile environments, and which are therefore likely to survive most traumas. (At the time of writing it was announced that 25-million-year-old bacteria had been brought to life after having existed in the gut of a bee trapped in amber. Bacteria have the remarkable ability to cover themselves in a layer of protein and become spores, which allows them to survive in a state of suspended animation.) The good news is that comet Swift-Tuttle is predicted to miss by 14 days, provided that no subtle gravitational perturbation in its orbit redirects it to come closer.

It is fairly safe to predict that starting around A.D. 2120 astronomers will be spending a great deal of time searching for this comet before it comes too close to the earth. I wonder how people living then will react to the knowledge that they are about to experience a close encounter with the most dangerous object in the solar system. Or will another one have hit by then, thus rendering the question moot? (To readers in the future: Is your government keeping a close eye on the sky? Are you prepared to take evasive action? What do you plan to do? Have you spread throughout the solar system to obtain as much advance warning as possible about the approach of a rogue comet, so as to maximize the chance of survival for the species? Or are you back in a stone age, books unread, because an impact restarted the clock of civilization?)

There is no longer much doubt about our precarious place in space. A clear message has been revealed by the threat of comet and asteroid impacts. The earth was not designed with the peace of mind in human beings as a specification. We find ourselves living in a dangerous environment, at least as seen from the perspective of long periods of time. We are perpetually poised on the edge of extinction and have been very lucky to get this far. How much longer will our luck hold?

10

RECONSTRUCTING

THE CRIME

JUST what happened to the dinosaurs? In the mind's eye, travel back to
the Cretaceous period, 65 million years
ago. First, land in a region of the world that will someday be called Oklahoma.
You are in the era of dinosaurs, although there are no longer as many species
about, worldwide, as there were ten million or so years before. In all, 23 species
roam their individual parts of the planet. It is their lack of spatial diversity that
will make them vulnerable to the catastrophe that is about to befall the earth.

So imagine you are there, together with triceratops, stegosaurus, velocirap-
tors, and tyrannosaurus rex. Mostly they live off the land, and some of them live
off each other.

On this day none of the animals on earth can possibly have any awareness
that they are about to disappear. Such a luxury will only be granted to a conscious species that has learned to explore the universe. For those who survive the
initial impact explosion and its immediate consequences, the coming months
will mark a terrible example of one of Cuvier's "brief periods of terror." In rapid
succession, all life will be subject to a holocaust of staggering proportion, horrendous blast waves, searing winds, showers of molten matter from the sky,
earthquakes, a terrible darkness that will cut out sunlight for a year, and freezing
weather that will last a decade. The ozone layer will be destroyed, and acid rain
will make life intolerable for species that survived the first few months after the
impact.

You are there and you have been observing an odd phenomenon in the sky. For thousands of years a great comet has loomed, repeatedly lighting up the heavens with its glorious tail and then fading away to reappear a few years later. Long ago it was seen to break into fragments, each of which was a spectacular sight in its own right. Sometimes one of those fragments seemed to loom ever so close to the earth.

For thousands of years, spectacular meteor showers have been seen whenever the earth passed through the tail of one of those comets, and sometimes dust drifted down into the atmosphere and disturbed the climate. There was even a hint that those encounters brought disease from space. But who could be sure? If you had been there earlier, perhaps you could have analyzed the nature of the stuff that drifted down and you wouldn't have been surprised to find that it contained the seeds of life, amino acids, and even hints of bacteria.

There were times not long before that showers of fireballs slammed into the planet. There were times when it seemed that the head of a comet filled the heavens, its blazing tail cutting out light from the stars beyond. Sometimes bright fireballs ended in loud explosions that were said to have devastated areas of the countryside beyond the horizon.

As you stand there, in that lonely place 65 million years ago, reflect on the fact that soon the face of the planet will be changed forever. Groups of dinosaurs will find themselves in the wrong place at the wrong time, no matter where they roam the earth, and all because a great, big ball of ice and rock is headed toward the planet on a collision course.

On this day you watch the increasing fireball activity, and the enormous fan of light that is the comet's tail. It seems to be growing larger. It is so bright and so close that it is visible by day and by night. As you stare, you notice something odd. If you look very closely, it is almost as if you can see a small, dark spot in the comet's nucleus, and it seems to be growing larger.

As startled as any of the creatures around you, your instincts tell you that there is something very wrong. Animals howl and seek shelter.

Now transport yourself a thousand miles to the south, out to sea, floating above the region of the world that will someday shape the Yucatan. Above you, the approaching comet looms. You arch your arm above your head to shield your eyes from the sun's glare. When you first noticed it, the object appeared a little larger than the moon but it is growing in size.

Thirty seconds tick by and the looming mass seems to be about three times the moon's diameter. Ten second pass, fifteen, then twenty. You reach out an arm as if to shield yourself and you notice that your hand can barely hide the object from view.

Five more seconds and you feel suspended in time, watching the onrushing monster in slow motion. You crouch in terror, two hands outstretched, barely able to cover the apparition as it begins to glow a fiery red. Within a second it bursts into incandescent fury and its glare fills half the sky with flaming light as hot as the sun. The sea begins to boil as the object passes into and through the atmosphere, puncturing a hole its own width, and smashes into the sea, gouging a hole kilometers deep beneath the sea floor.

Enormous masses of water blast outward, vast landslides are triggered beneath the sea, and the resulting tsunamis literally empty the basin that is the primal Caribbean while the region deep down below the earth's crust shudders and rebounds and sends a splash of molten rock upward. A crater 20 kilometers deep and 180 kilometers across is gouged into the sea floor as the earth shudders and seismic shocks radiate outward at thousands of kilometers per hour.

A fireball bursts outward and upward and spreads rapidly through the stratosphere to encircle the planet. Trillions of tons of rock have been vaporized and blasted into space. The mushrooming plume of material launched by the impact recondenses in space and falls back so that the planet is surrounded by a billion, trillion "shooting stars," a flaming shower of particles that burn up in the atmosphere on their way down. The sky becomes so hot that it feels like you are inside an oven set to broil. There is so much material up there that it is as if the sky itself has caught fire. Creatures everywhere are literally broiled to death, vegetation bursts into flames, and great coniferous forests are torched into incandescence. This happens all around the globe, not just near the impact site.

The comet has been blown to smithereens and its material joins billions of tons of sulfur-laden dust, some of it from the comet itself, some of it from the sulfur-rich rocks that lay beneath the impact point. During the next few years the sulfurous stuff mixes with water to make acid rain. The acidity grows worse as nitrogen oxides created in the heat of impact add to the chemical brew that produces a corrosive rain that will strip vegetation bare and run off into the oceans to kill off life along the continental shelves.

At the moment of impact, a frightful sound resounds through the atmosphere as air rushes in to fill the vacuum created by the passage of the comet's nucleus. A thunderclap of indescribable proportions booms out over the hemisphere, a blast wave traveling with such ferocity that it flattens everything in its path for a thousand kilometers in all directions.

Within hundreds of kilometers of the impact, droplets of molten rock splash into space where they cool before falling back through the turbulent atmosphere to rain back to earth. They will later be identified as tektites and their strewn field surrounds the bull's eye where the comet disintegrated and dug a crater.

Within an hour of impact the rumble of the earth is heard around the world and earthquakes toss everything into the air. With magnitude 12 to 13 on the Richter scale, the earthquake mangles solid rock as the ground buckles.

All around the planet the seismic shock rumbles. As it travels its energy begins to focus so that at the antipodes it gathers and the planet's surface buckles and heaves 20 meters. Cracks develop and lava flows begin, which, in turn, release sulfur dioxide and carbon dioxide and chlorine to poison the atmosphere further, an atmosphere that responds to the injection of energy by heaving violently and triggering enormous storms that tear at ever corner of the planet. Out of some of those storms water pours and douses fires.

Back were you first stood in imagination, where everything was torched and blown away as the ground buckled, your ghost sees a thunderous wave approach that washes the world clean. This is the tsunami created by the gargantuan splash where the comet struck the sea to send the wave rippling and thundering across the water and onto land, depositing thick layers of silt over a vast area that would some day be shaped into Mexico, the Caribbean Islands, and the southern United States.

Eight hundred kilometers from the impact a tsunami more than a kilometer high washes over the North American continent to create ripples in the land that will be preserved and etched in geological strata for 65 million years to come. Those ripples will be mute testimony to the story of this cataclysm. A hundred meters of deposits dragged from the bottom of the sea cover the islands and the coastal regions of the main land and boulders the size of automobiles land 500 kilometers from the impact (in a country later to be called Belize).

After the fire, the wind, the buckling earth, and the tsunami, a vast cloud of soot and dust envelopes the planet, cutting off all sunlight, everywhere. Utter darkness descends upon the shattered planet and lasts for half a year, a darkness so total that no sunlight penetrates. Plants die from lack of photosynthesis and herbivores not already wiped out in the aftermath, starve or freeze to death.

The cataclysm is no greater than countless such impacts that shaped the earth in its formative years. During the first few hundred million years of the planet's existence, impacts with giant comets were violent enough to evaporate the waters of the oceans and sterilize the planet. Back then life was, at best, in the form of primitive bacteria. Now, however, the earth's biosphere is teeming with creatures from bacteria to dinosaurs, from algae to giant trees, and from plankton to enormous sharks.

The food chain has been destroyed but there are those creatures which revel in the catastrophe—ratlike species and cockroaches that feed on detritus flourish.

In the awful aftermath of impact, global extinction results. Some of the lucky creatures that survived—and some of them managed to do so even within sight of the impact—were those that lived under the ground. They were sheltered from the searing fire from the sky, the winds, and the devastating tsunami, which might have done no more than rush over them as they hid in fear. Even the buckling earth could do no more than temporarily destroy their burrows. When the great cold descended upon the earth in the months to come, they ate the roots of plants and trees that had been trimmed to ground level.

The burrowing creatures survived the extended period of cold that gripped the planet in freezing temperatures for a decade. They survived the rain that fell as strong as battery acid to cut swaths of death through the rivers and into the oceans to wipe out vulnerable marine life along the continental shelves, species that survived the original heat and winds and waves. As a result, three-quarters of marine species went extinct along with the dinosaurs, as well as oh-so-many creatures and plants that inhabited the planet.

Even as these terrible changes were wrought on the earth's surface, high in the atmosphere, in the blanket of soot and smoke and dust, the heat from the original fireball brought together nitrogen and oxygen and for each megaton of explosive power, thousands of tons of nitrogen oxides (NO) were produced. Each ton of NO destroys a hundred times as much ozone. Within days there was no ozone left to protect the planet from the sun's damaging ultraviolet radiation. But for a while that wouldn't matter much, because the dust and soot blanket did a perfectly adequate job of providing protection.

There was so much carbon dioxide in the atmosphere that when the clouds cleared, and even after the ozone layer struggled back into existence, after 20 to 30 years, the earth experienced a greenhouse effect that initiated a period of prolonged warmth.

The echoes of the collision resonated in the environment for centuries and millennia as new species emerged from the chaos that was precipitated by the impact, and as the climate oscillated until it settled down into a more "normal" state.

If you could have watched at the point of impact, you might even have been fascinated by the formation of the crater. The impacting comet stopped in its own width and literally turned inside out as the pressure of several million atmospheres and shock temperatures of many tens of thousands of degrees transformed the comet material into a ball of hot gas, and as the ground into which it crashed behaved like a liquid. The crater rim rose to several kilometers high as the splash developed and its center was immediately filled with molten rock— lava. The ejected material splashed out and spread molten material that formed tektites, which fell over an area to distances of hundreds of kilometers.

As regards the likelihood of survival at the time of the K/T impact, any creature above the surface within about 5,000 kilometers would not live. If you weren't roasted, killed by the pressure of the blast, blown away, cooked in the hot winds, or dashed to death by the earthquake ripples, drowned in enormous tsunamis or the floods produced by torrential acid rain, you'd soon perish of starvation or freezing. Whether you survived would be just a matter of luck. Were you in the right place at the right time? In a cave perhaps, when the winds came? In a valley when the firestorm blew by overhead? Did you have food and shelter when the acid rain fell, or later when the great freeze began?

There was one safe place from which to watch the impact: the moon! If we want to maximize the chances for the human species to be around for a very long time we must spread into the solar system.

Our curiosity about the ways of nature have lead to a sobering discovery; that the life we know and enjoy in the twentieth century could be snuffed out in an instant. It surely will require more than scientists simulating the catastrophe in computers to convince us that this is so. That is why many of those same scientists are beginning to look very closely at further evidence that may be dug up from around the crater in the Yucatan, the marker of the impact that has become the center of fascination for all catastrophists. In fact, it may the single most dramatic indicator that tells the human race that life on earth is not the way we thought it was, that instead our origin and evolution have been mediated by cosmic collisions.

The big question regarding the K/T impact still concerns just what hit the earth. Was it an asteroid or a comet? Some scientists have argued, based on the spreading of 250,000 tons of iridium as well as the energy estimated to be needed to produce the crater, that a comet is implied. An asteroid is still a suspect, but only if most of the iridium escapes earth, or is still in the crater's melt rock (the molten material that flowed to the center of the crater as soon as it was formed), something that will be determined through further drilling. The melt rock samples that were collected more than a decade or so ago suggest that much of the iridium is still down there, deep beneath the ejecta blanket that covers the crater and the surrounding area reaching to Texas.

This limited description about what happened 65 million years ago when a comet slammed into the earth is based on reports of computer simulations of the consequence of cosmic impact. This has become a tremendously vigorous area of research in which better simulations quickly render old ones obsolete. This means that the details referred to here will rapidly change, but their essence remains constant. The K/T impact was a frightful happening, and the stunning thing is that the computer simulations have recently been tested by comparison with observations of the train of the comet that slammed into Jupiter in 1994. If anything, the models may have erred on the conservative side.

Despite the terrible implications of comet impact that are implied by the physics and chemistry of the phenomenon, there are those who claim that the notion that the species went extinct because of a comet impact is only a hypothesis, one of several that seem likely. That may have been true 15 years ago, but the data now point overwhelmingly toward an impact, not to unassisted volcanic eruptions or even to slow climate change, as the trigger for the K/T extinctions. The impact idea is no longer an hypothesis; it is a theory that is being tested by many, many scientists. (In the 15 years since the Alvarez study was published, an average of over 200 scientific papers a year have been written on impacts and mass extinctions.)

To suggest that the dinosaurs were not wiped out by the comet impact is something like saying that the 1,500 victims of the sinking of the Titanic in 1912 did not die in the collision with an iceberg. Technically, that is correct. No

TABLE 10–1 Suggested mechanisms for K/T extinctions

Agent	Mechanism	Time scale [a]	Geographic scale [b]
Dust in atmosphere	Cooling	Y	G
	Cessation of photosynthesis	M	G
	Loss of vision	M	G
Fires	Burning/broiling	M	G
	Soot cooling	M	G
	Pyrotoxins (chemicals from fires)	M	G
	Acid rain	M	G
NO_x generation	Ozone loss	Y	G
	Acid rain	M	R
	Cooling	Y	G
Shock wave	Mechanical pressure	I	R
Tsunamis	Drowning	I	R
Heavy metals in air and rain	Poisoning	Y	G
Water/CO_2 in air	Warming	D	G
SO_2 injections	Cooling	Y	G
	Acid rain	Y	G

a Y = years; M = months; D = days; I = immediately.
b G = global; R = regional.

one died in the collision. Everyone who died did so later, as the result of drowning, hypothermia, heart attacks, falling to their death, or from being crushed by swinging lifeboats. In the same way, virtually no species went extinct at the moment of impact. It is what happened afterward that took its toll. That was when habitat was destroyed and entire species disappeared because they burned, drowned, froze, or starved to death. Some of those that survived this horrible list may have been wiped out later by overexposure to the solar ultraviolet light that poured down onto the planet after the clouds cleared. Poisonous fumes from volcanoes triggered by the impact also took their toll, and long-term climate change that acted over subsequent centuries may also have done its share.

A group of scientists lead by Owen Toon of the NASA Ames Research Center made a study of the mechanisms that could have contributed to the K/T extinctions. These are summed up in Table 10-1, taken from their report.

One entry in the table is worth attention before we leave this drama. When in the aftermath the world is surrounded by dense dust and soot clouds, no sunlight is let in for months. That means that no creature can see anything and those that depend on eyesight to find food surely die. It also seems likely that those that depend on a sense of smell must have had a very rough time, because the atmosphere was so polluted with soot and smoke and nitrogen oxide and sulfur dioxide. It is a wonder that any species survived at all. Some did. After the dust settled, they emerged to spread over the newly shaped world and multiplied and found new niches in which to fulfill their destinies. Today, 65 million years later, scientists working within many disciplines have figured out what happened on that remarkable day when a comet slammed into the planet and changed the course of evolution.

11
DEATH STAR OR COHERENT CATASTROPHISM?

*O*UR instinct for survival drives us to learn as much as possible about what goes on around us. The better we understand nature, the better we will be able to predict its vagaries so as to avoid life-threatening situations. Unfortunately, nature is seldom so kind as to arrange for disasters to occur like clockwork, yet that does not dampen our enthusiasm when even a hint of periodicity in a complex phenomenon is spotted. This helps account for the furor that was created when a few paleontologists claimed that mass extinctions of species seemed to recur in a regular manner. A cycle, a periodicity, had been found! That implied that perhaps they might be able to predict nature's next move.

This is how I interpret the extraordinary public interest that was generated by the claims made around 1984 that the mass extinction phenomenon showed a roughly 30-million-year period (others said it was 26 million years). Almost immediately, several books appeared on the subject as well as many, many articles in the popular press and in science magazines. This activity marked the short life of the Death Star fiasco.

Given our instinctual urge to look for order in the chaos of existence, the identification of a periodicity in mass-extinction events was a great discovery, if real. What was not highlighted by those who climbed aboard the bandwagon, however, was that the last peak in the pattern occurred about 13 million years ago. If impact-related mass extinction events were produced every 30 million years, there obviously was no cause for concern that we would be hit by a 10-kilometer object in the next 17 million years. Phew!

I think that the suggestion that mass extinctions occurred on a regular cycle caused as much interest as it did because we all want to *believe* that there is no immediate danger to us.

The Death Star fiasco began when David Raup and John Sepkowski of the University of Chicago published a report claiming that mass extinction events recurred about every 26 million years. They were followed by Michael Rampino and Richard Stothers of the Goddard Institute for Space Studies in New York who claimed that the period was more like 30 million years, at least during the last 250 million years. They also claimed that terrestrial impact craters showed an age distribution that suggested a cycle of 31 million years. These periodicities are not so obvious that anyone looking at the data would immediately agree. The phenomenon has to be demonstrated in a statistical manner, which made me very skeptical.

Rampino and Stothers suggested that the basic periodicity is caused by the sun's motion in and out of the disk of the Milky Way galaxy (Figure 11-1). While passing through the disk, the Oort comet cloud would be disturbed by the Galaxy's tidal influence so that more than the usual number of comets would visit the inner solar system, and thus pose a greater threat to the earth. This might happen, and it had already been suggested by Clube and Napier, whose work continued to be ignored in the United States.

A more detailed variation on this theme suggested that when the solar system runs into and through a cloud of gas and dust between the stars, a so-called interstellar cloud, its gravitational effect would do the trick and kick comets toward the sun. Astronomers in-the-know pointed out that there are not enough so-called giant molecular clouds to create a reasonable probability of running into them in the required period of time. Some proponents then talked of a solar system encounter with more "normal" interstellar clouds, not those filled with lots of dust and complex molecules, but instead with mostly hydrogen gas. According to my own research, such "clouds" do not even exist! They are fictions created by insufficient data combined with theoretical speculations about the nature of interstellar matter. My research, as well as that of several colleagues, has shown that normal interstellar structure is highly filamentary and that discrete, dense, medium-sized "clouds," of the type invoked by Rampino and Stothers for producing comet showers, just don't exist.

Three British astronomers expressed their skepticism about galactic modulation far more explicitly. M. E. Bailey of the University of Manchester together

Figure 11-1 The spiral galaxy NGC 2997, which shows structure not unlike that thought to exist within the Milky Way. The bright lanes indicate where star formation has recently occurred and where the bright young stars now reside. A star passing through the bright spiral arms, marked by dark, narrow dust lanes on their inside edge, would feel the influence of the gravitational pull of the arm, which might perturb the comet cloud surrounding the star. This phenomenon is thought by some to generate a periodicity in comet collisions with earth as the sun passes through similar spiral arms in the Milky Way. (Courtesy Anglo-Australian Observatory)

with D. A. Wilkinson and A. W. Wolfendale of the University of Durham showed that, when you really get down to it, there is no credible astronomically induced process that can be imagined to be the primary cause of comet showers with periods of 30 million years or so.

As an alternative to the galactic "clouds" and the Galaxy hypotheses, the most popular explanation for comet showers, in the eyes of journalists and reporters at least, was the Death Star hypothesis, a nonsensical idea from the start.

The Death Star, which some called Nemesis, is supposed to be in a long-period orbit about he sun. Every 30 million years it is believed to come close to the sun and pass through or near the Oort comet cloud. Comets are then likely to be propelled into new orbits into the inner solar system. When Nemesis is near, things become stirred up so badly that the chance of earth being struck by a rogue comet is supposed to go way up. By implication, impacts would also be more frequent on the moon and the other planets and their moons. It is not obvious that they were. Dating lunar craters would be an interesting experiment for future colonists, where dating is used to indicate the process of estimating ages of the lunar rocks.

Based on questions posed after lectures I have given on the topic of impacts, it appears more people have heard of the Death Star hypothesis than any other aspect of the dinosaur extinction story. Why is that? The answer, I think, can be found in the psychological attraction of picturing something dark, mysterious, and invisible lurking out in space. In addition, this unknown thing poses a threat to us. This is the stuff of horror fiction. No one really believes this model anymore, but it did cause quite a few publishers to beg people to write books about the subject. What more did one need to capture the public imagination? The Death Star "thing" had the power to influence the destiny of life on earth, and fortunately it was so far away from us at this time that we were not immediately threatened, and wouldn't be for millions of years to come.

The problem with the Death Star is that, in order to spend 30 million years going around the sun, it has to move a long way into space before coming back. It needs to move a couple of light years away according to Daniel Whitmore of the University of Southwestern Louisiana and Albert Jackson of Computer Sciences Corporation of Houston. A similar calculation by Marc Davis and Richard Muller of the University of California and Piet Hut of Princeton suggested that the Death Star's maximum distance from the sun would be 2.4 light years. They added that it therefore posed no threat for the next 15 million years.

The flaw in the Nemesis theory is that any star in an orbit about the sun that takes it so far from home would not last long in any orbit. It would be gravitationally lured by other stars to join them for a while instead. In the solar neighborhood stars are separated by an average of 8 light years or so. Sometimes they are closer, at other times farther apart, and when the sun is in the disk of the Milky Way there would be a lot more stars around to lure Nemesis out of the sun's clutches. It would soon lose touch with the sun, after a few orbits at most, and be lost.

Despite this fatal flaw, the Death Star hypothesis hit the headlines while most astronomers all but ignored the idea. They also scoffed at a variation on the theme that invoked a tenth planet to do the dirty work. The idea of a tenth planet has been around for a long time, and the quest to find it has lured many an astronomer to performing fruitless searches of the heavens. This hypothesis invokes the existence of planet X to disturb the Oort comet cloud from time to time, but it seems to be about as relevant as the Death Star idea.

At this stage in our narrative the reader will be asking how, then, the 30-million-year periodicity can be explained. My response is that I have found the claims for periodicity utterly unconvincing.

Richard Grieve of the Geological Survey of Canada is a widely recognized authority on terrestrial cratering and thinks that, in view of the uncertainties in specific crater ages, little or no evidence for a period in their formation can be trusted. This leaves only the mass-extinction data to support claims of periodicity, and one has to be fond of statistics to be even tempted. A great deal more information will be needed before we can trust such claims. In the meantime, we have to face a more sobering conclusion, that mass extinctions have no underlying period and may happen at any time (see chapter 13).

Even without a periodic bombardment of the earth, the role of comet showers in triggering mass extinction may nevertheless be crucial. Piet Hut and his colleagues pointed out that many of the extinction events seemed to occur in a stepwise fashion, which implied multiple impacts. They concluded that "astronomical, geological and paleontological evidence is consistent with a causal connection between comet showers, clusters of impact events and stepwise mass extinctions, but it is too early to tell how pervasive this relationship may be." That was said in 1987. Their work reinforced an idea that seems to have persisted since then, which is that the mass extinctions and cratering are not periodic, but *clustered*. In other words, when a large impact occurs it tends to be a member of a family of such visitations, even if they are spread over millions of years.

This scenario received interesting confirmation by the 1994 report by Steven Stanley of Johns Hopkins University and Xianagning Yang of Nanjing University. They found that the infamous Permian mass extinction of 245 million years ago, which saw the disappearance of 95 percent of species, involved two events separated by 5 million years. They studied single-celled organisms called fusulincean foraminifera (also known as forams), which are abundant in the fossil record for 100 million years and then disappeared at the Permian extinction. The first impact wiped out 71 percent of marine species and the second cleared out 80 percent of those that survived the first. This means that here two major impact events were separated by as little as 5 million years.

Despite the enthusiasm engendered by the early claims of periodicity in mass-extinction events, there is no way we can rest assured that a serious comet crash will not occur for millions of years to come. We can only find succor in the belief—and it is a matter of belief—that very large impact events occur infrequently. This is the point of view hawked by the U.S. experts on the subject. The same cannot be said for smaller events that could hasten the end of civilization. Such events will happen often, and we may even now be in a period of relatively great danger.

David Asher, also from Oxford University, and Duncan Steel have teamed up with Bill Napier and Victor Clube to produce what may be one of the most important interpretations of the comet danger that has yet been published. For the sake of brevity I will refer to this quartet as the Musketeers, partially to recognize their independence of mind. They acknowledge that near-earth asteroids (NEAs) are a threat to our planet, but suspect that the introduction of giant comets (more than 100 kilometers across) into the solar system, some of which break up while in earth-crossing orbits, is a more likely and dangerous phenomenon.

Based on their studies of the observations of comets, meteor streams and near-earth asteroids, as well as computer simulations of the phenomenon, they think that every 20,000 to 200,000 years a giant comet enters the solar system to pass closer to the sun than the earth. Through the stress of passing close to the sun, collisions in the asteroid belt and interactions with Jupiter's strong gravitational pull, it fragments into many smaller objects. The influence of Jupiter continues and the orbits of the fragments evolve into a stream of disintegrating material that may pose a threat to our planet for as long as 50,000 years. This is something like what happened to the Kreutz comet family described in chapter 4; these are not in earth-crossing orbits but they have been disintegrating for thousands of years.

What makes this danger more insidious is that a stream of cometary material in an earth-crossing orbit will suffer a wobble known as precession. This means that the stream might only intersect the earth's orbit every few thousand years. Then, for a century or so there will be large numbers of impacts every few years.

Many comets have been seen to split when they come too close to the sun; Figure 4-3 shows the comet of 1844, which broke into six distinct objects. Comet Biela is another example. On January 13, 1846, it broke into two and at its next return, in 1852, the two components were separated by 2.4 million kilometers. When they were due back in 1865, two orbits later, all traces of the fragments had disappeared. But in 1872 the earth ran into an extraordinarily vivid meteor shower, produced by comet Biela's remnants. At one site it was estimated that over 160,000 shooting stars were seen in an hour. Today the debris of comet Biela creates the Andromedid meteor shower. The meteoroid stream that produces the shower is filled with mostly very small particles, but it grew out of two large objects that broke up and dispersed stuff along their orbits until their entire paths through space became filled with debris. When earth runs in the stream, only the small particles collide with our planet. Had the breakup occurred at a different time, the earth could just as well have run into some of the larger fragments with devastating consequences.

Not long ago the earth ran into the debris of another shattered comet. The rendezvous occurred in June 1975, but the focus of the impacts was the moon, which was struck by many one-ton boulders over a five-day period. That was seen in data from the seismographs left by the Apollo astronauts. The meteor stream in question is called the Beta Taurids, a daylight shower that, according to Clube and Napier, harbors the greatest danger to life on earth. Again, if the earth had encountered that stream earlier in its history, it would have been struck some hefty blows. As Clube and Napier put it, it is in the aftermath of the initial splitting or fragmentation of a giant comet that the greatest danger will be posed to our planet. The danger could persist for hundreds if not thousands of years as a concentrated population of boulders and larger bodies remains in a swarm that marks where once was a comet's orbit. The chance of catastrophic impacts decreases as the swarm spreads out and the fragments continue to disintegrate even further.

Clube and Napier criticize scientists in the United States for being too enamored with the single, rare impact scenario, essentially a uniformitarian point view that pictures long periods of quiet with an occasional setback superimposed. Instead, we may have to consider that impacts are far more frequent and even regular, albeit smaller. These impacts may not cause extinction of species but will do nasty things to civilized societies unprepared to deal with climate disruptions of the type that may accompany such events. Our civilization would not survive an impact winter that cut out sunlight for several months. Crops would immediately die because of lack of photosynthesis, and the food chain would collapse. Global famine would result. Therein lies the greatest danger from a small impact.

It is unlikely that the earth will ever be struck by a large 100-kilometer-sized object. It is far more likely that we will suffer from the rain of debris created by the breakup of a large comet that leaves the solar system littered for tens of thousands of years. The breakup of a 100-kilometer comet could produce a million objects, each one kilometer in size. This is the essence of the story offered by the four astronomers who claim that about 20,000 years ago something like this happened.

The Musketeers suggest that long-term periodicities that may or may not exist in the mass-extinction events are not the issue, and that collision with a rogue comet is even less relevant. Instead, the greatest danger to civilization may be posed by objects in what is called the Taurid Meteor Stream. The Taurid meteor shower occurs from December 10 to 12 each year and the daytime equivalent, the Beta Taurids, is seen in late June. These two showers are produced by small particles at the fringe of the stream. The Tunguska object was probably a larger fragment of the same stream.

Comet Encke, a small comet 5 kilometers across, is in an earth-crossing orbit and is a member of the Taurid Stream. It takes only 3.3 years to orbit the sun and is not visible to the naked eye. It may be the remnant of a larger parent whose disintegration produced this band of debris, which today may pose a greater threat to our planet than any other group of near-earth asteroids. Comet Encke may also have produced the zodiacal dust cloud, which, according to Clube and Napier, was once a lot more prominent than it is now.

According to the Musketeers, a larger Taurid Complex consists of at least four meteoroid streams (including the Taurid Stream) that produce regular meteor showers and as many as 8 other streams. It also includes at least 18 Apollo asteroids, 13 of which belong to a comet Encke group and the other 5 to the Hephaistos group, associated with an asteroid of that name. The two groups may represent the remnants of two comets that entered the inner solar system through the same Jupiter-mediated corridor. Many more members are likely to be found.

No one is yet sure what else the Taurid Stream contains, or where the bulk of its material might be concentrated. Its structure is complex and its existence is only just beginning to be recognized. This poses immediate problems for those who plan to take seriously the protection of our planet from future impact catastrophes.

In chapter 8 we considered Clube and Napier's remarkable suggestion that evidence for smaller and potentially civilization-destroying impacts can be found in the historical record. They have gone so far as to blame a lot of recent heaven-sent disasters on impacts by objects that are part of the Taurid Stream. The Musketeers have since argued that collisions with objects in the Taurid Stream produce episodes of major atmospheric detonations with important consequences for civilization on time scales of 10 to 10,000 years.

No one doubts that the meteoroid streams through which the earth passes every year had their origin in comet breakup. As the fragments disintegrated, they seeded the path of the original comet with, among other things, small dust particles that produce meteors or shooting stars. The orbits of those particles evolve through interactions with the planets, which causes the meteoroid swarm to exhibit spatial structure. Different parts of the original streams evolve in slightly different ways, changing their orbits so that they intersect the earth at different times of the year. Within each stream there are clusters of objects, which means that in some years we may see spectacular meteor storms and in other years barely any enhancement in the typical numbers of meteors expected on any given night.

The frequency of meteor activity, and the likelihood of running into a large fragment that still shares the meteors' orbit, varies in yet another way. The orbits of the fragment, whether dust, rocks, or boulders, or even small comets that are very difficult to spot, begin to precess, which means that they wobble slowly. This, in turn, means that their paths do not always intersect the plane of the planetary orbits (the ecliptic) in the vicinity of the earth. Thus a cycle begins to manifest. The wobble of precession allows the meteoroid stream to sometimes cross the ecliptic inside the earth's orbit and often outside it. When it crosses the ecliptic at the earth's orbit just when the earth is in the neighborhood, things can get dangerous. If you add to that the existence of clumps of material in the streams, because the disintegrating material is not yet fully dispersed, then occasionally the planet has a very high probability of being struck repeatedly over a relatively short period of time, lasting a century or so, by many small objects in the 50- to 300-meter-size range.

The Musketeers think that this is what happens every few thousand years, which ties in with what Clube and Napier were saying for nearly two decades based on their study of historical records. What the Musketeers have been doing recently is to gather data to back up their assertions, data regarding which objects are members of the Taurid Stream, for example. Their model goes by the name "coherent catastrophism." Duncan Steel has defined it in this way: "Cataclysms visit wide areas of the planet due to the coherent arrival of many impactors in a few days. It is entirely feasible that within those few days the earth could receive hundreds of blows like that of the Tunguska object."

Historical records from China provide evidence of fragmentation having occurred in the Taurid Stream. Those records tell of enhancements in fireball numbers that occurred at various times; for example around A.D. 1100 and 1000 B.C. The Musketeers hypothesize that a giant comet began to break up 20,000 years ago and that the larger fragments still move about together so that there may still be a large mess of stuff somewhere along the Taurid Stream orbit. Unfortunately, no one knows where such clusters may be located and none of the present search programs are designed to find them. Fortunately, however, we do not run into a cluster of large debris every year, but because of precession it appears that we will do so every few thousand years, if past records are anything to go by. The proof will lie in finding evidence that such a cluster of asteroids actually exists, no small task, given how little is known about near-earth objects (NEOs), whether comets or asteroids. But then, according to the Musketeers, families of asteroid orbits have been found to be associated with the Taurid Stream, so what more does one require?

Meteors associated with the Taurid Stream are clearly bunched. The brighter meteors are seen in showers, whereas the fainter ones, detected by means of radar, are more uniformly spread around the sky. This means the larger ones are in a narrower stream and the smaller ones more dispersed in space. Most "normal" meteor showers are seen annually, so the meteor sized particles are spread out along the full path of the orbit. According to computer simulations of their dispersal rates, they should lose all identity with their original comet orbits within 100 to 1000 years. However, the larger particles remain clustered close to the original nucleus, as for example for the Leonid meteors associated with comet Temple-Tuttle, which shows a 33-year cycle of meteor *storms*, particularly vivid meteor showers.

Cyclic behavior of the meteor showers, storms, and impacts is thus to be expected, but these are short-period cycles tied to the regular motion of the earth around the sun, and offer us no consolation for future security. As regards the possibility of serious impacts, those depend only on the size of the stuff that still exists in the clusters of material that may lie in the orbit of a given stream.

The Taurid Stream is known to contain many boulder-sized objects, some of which were involved in a remarkable string of impacts on the moon referred to before, which occurred in late June 1975 as detected by lunar seismographs.

They recorded a series of small collisions that set the moon "ringing." The instinct to have left seismographs there when the astronauts returned to earth may yet save civilization, at least if you allow for the warning they give about enhanced bombardments from space at certain times of a year.

The Musketeers are acutely aware that their model has yet to be accepted by the community of (mostly U.S.) scientists involved in the study of NEOs. As they wrote, "The existence of kilometer-sized asteroids in the Taurid Complex is apparently doubted by some workers in the field of minor solar system bodies." Here one is reminded of the initial reaction on the part of the majority of scientists to the original suggestion by the Alvarez team that an asteroid impact was the trigger for the K/T extinction event. It is very difficult to accept a new idea, even when one understands what it is.

They also point out that the doubt referred to here has never been substantiated in the scientific literature. Instead, the idea has simply been ignored by U.S. "authorities" on the impact-extinction debates. To ignore coherent catastrophism is to take a huge risk, because in the relatively short-term our future survival may be depend on whether this model is correct or not. If it is, then we have a great number of important things to consider. It implies that impacts capable of regional catastrophes are not infrequent, random events that recur every 100,000 to a million years. Instead, the stormy visits associated with the earth running into the crowded parts of the Taurid Stream, which occur every few thousand years, are what we should worry about. The subsequent impacts may not produce a global extinction catastrophe, but are likely to produce a serious disruption to modern civilization.

According to the Musketeers, we are now in a lull between impacts related to the Taurid Stream. We may even be in for a thousand years of grace, but then the skies will again become hazardous. Even this outlook should offer us scant consolation, because we will still have to contend with unexpected arrivals of rogue rocks from the Taurid Stream, or from the larger Taurid Complex of streams, or small comets that arrive totally uninvited.

We are apparently confronted by a vast gulf between two points of view within the community of planetary scientists discussing the likelihood of impacts violent enough to destroy civilization. On the one hand, there are those who hold an essentially uniformitarian point of view, which pictures relative quiet between lone impacts by large objects separated by tens to hundreds of millions of years. The proponents assume that we know enough about the earth-approaching bodies to tell about the long-term nature of this population, and hence of impact events from single visitors from space, a version of catastrophism based on uniformity. They do not allow for large fluctuations in the impactor population. This point of view offers consolation as regards the likelihood of impact. After all, virtually everyone agrees that a direct hit by a lone comet is very unlikely, and that such events are surely separated by millions of years.

The coherent catastrophism point of view offered by the Musketeers is less comforting. Catastrophic impacts are not random. Instead, the earth occasionally finds itself in a hazardous environment resulting from the past breakup of a large comet that spawned many, many smaller objects, collision with any one of which may trigger the end of civilization.

Another aspect of the more immediate danger has been highlighted by Mark Bailey, of Liverpool John Moores University, and his colleagues who have said that if we wander into regions of old streams where there is a great deal of dust, the earth's climate will change and this, too, poses a distinct hazard to civilization. Evidence that the zodiacal cloud has undergone profound changes in historical times is very strong. Satellite studies of the detailed structure in the cloud show bands of material that represent different orbital evolution of the debris blown out of a close pass of a comet something like 20,000 years ago.

Coherent catastrophism paints an image of short periods of greatly enhanced impact activity spread over a few centuries with such episodes separated by thousands of years following the breakup of a large comet. Such breakups may occur every few tens of thousands of years. This image offers both good news and some bad news. The bad news is that we may currently be surrounded by the debris of a giant comet that broke up about 20,000 years ago. The good news is that we are in a lull between storms. The evidence presented by known meteoroid streams, the Taurid Complex of streams in particular, as well as the orbits of Apollo asteroids and a few nearby comets in short-period orbits, point to this phenomenon. However, there may be more meteoroid streams out there we do not yet know about. This point of view also reminds us that all the meteor showers (Table 3-1) tell of the past breakup of large comets, and that perhaps we are never entirely free of the hazard.

Kenneth Hsü of the Swiss Federal Institute of Technology, in commenting about how long it has taken the earth sciences to accept the ideas of catastrophism over uniformitarianism, which has held center stage for nearly two centuries, said that "Substantive uniformitarianism [the new name for the old beliefs] is the epitome of the ignorance and arrogance of nineteenth century scientists." They believed that the world was a benign place and that you can account for what happened a long time ago by comparison with the sky as it appears today, or at least in terms of phenomena that can be observed in our lifetimes. Duncan Steel thinks that there is little to add to Hsü's sentence except a century, and that many scientists continue to be locked in the struggle between catastrophism and uniformitarianism.

In writing about uniformitarianism versus catastrophism in the K/T extinction debate, Hsü added that the nineteenth-century scientists were "sufficiently arrogant to believe that what is not observable cannot have happened, forgetting the brevity of human life compared to the course of earth history." He felt that the K/T debate is likely to continue "until the present generation of antagonists has met its fate." That may be true for the K/T debate, but the tragedy for our species may be that a similar debate is beginning to shape up as regards the probability of future impacts, and this argument may only be settled when the next impact injects certainty into the estimates.

The coherent catastrophist point of view begs us to pay more attention to claims made by those who have examined the historical record. The proponents go so far as to suggest that the last ice age was brought to an end by an impact, and that significant impact events stamped their mark on the evolution of societies in the past several thousand years. When you begin to look more closely at this point of view, it appears that we have been lucky to get this far.

The debate that has begun as regards coherent catastrophism versus the new version of uniformitarianism is of more than intellectual interest. The way it is settled will determine how we regard our future, even whether our species has a future.

The notion of coherent catastrophism has yet to sweep U.S. astronomers before it. Uniformitarianism is still deeply rooted within this community. To accept the picture of coherent catastrophism is to admit that civilization may dwell under a greater threat than we are prepared to admit. I also suspect that many modern catastrophists are hesitant to grasp the full implication of catastrophes that may yet befall our planet. It is "politically correct" to act more conservatively than one's scientific colleagues, yet we may have to confront some radical ideas if we wish to assure the survival of civilization.

Those who first recognized the short-term threat implied by coherent catastrophism are struggling to make known their view, because their theory also implies distinctly different search strategies to find the rogue comets and killer asteroids. The near-earth object searches so far (chapter 16) have concentrated on objects approaching us from the asteroid belt, objects that move along the plane of the planetary orbits, the ecliptic. But giant comets that from time to time visit us, and the streams of debris they create when they break up, pay little attention to the ecliptic. They can arrive from any angle, which means that fragments spawned 10,000 years ago may yet approach from directions well away from the ecliptic. That is why the Musketeers argue that we should spend a great deal more time on searching for the remnants of comet breakup rather that the lone asteroid on a collision course with earth.

12
CRATERS AND
TSUNAMIS

*U*NTIL the lunar explorations began in earnest in the 1960s, the Barringer crater in Arizona was believed to be one of the few, if not the only, impact crater on earth. Before the moon landings, many scientists thought that lunar craters were volcanic in origin and that the moon might be covered in a layer of volcanic dust meters thick so that astronauts would sink up to their eyeballs when disembarking from their space capsules. A pleasant sense of relief greeted the news that the first unmanned lunar spacecraft did not disappear into the dust.

For a century or more it was doubted that lunar craters were produced by impacts because it was assumed that such craters would seldom be circular. It seemed obvious that circular craters could only be produced by objects falling straight down, a rare situation, since meteorites are likely to approach from random directions, especially on the moon where there is no atmosphere to slow them down before impact.

W. M. Smart in 1928 stated this explicitly: "Objections to lunar craters being caused by meteors is that the craters are round and there is no *a priori* reason why meteors should fall vertically and in no other direction." He also shuddered at the notion that the impactors would have to be as large as asteroids to create the lunar basins. At about the same time, Thomas Chamberlin ruled out impacts on the moon because there was no evidence for an appropriate population of objects anywhere in the solar system that could have made the craters That was in 1928 when near-earth asteroids had not yet been found, and when little was known about the history of the moon or the formation of the solar system. Richard A. Proctor in 1896, however, had concluded that because so many meteors continued to fall to earth that the planet and the solar system were still forming. To him, this made more sense than to blame the formation of the planets on "the creative fiats of the Almighty." There is merit to his point of view, because today's bombardment merely represents a faint, ongoing manifestation of the process of accretion that assembled the planets in the first place.

Jay Melosh has written a scholarly summary of impact cratering wherein he tracks down the origin of the reluctance of scientists to accept that lunar craters were produced by impacts. In part the misconception rested on authoritative statements that lunar craters had to be volcanic, with William Herschel in 1787 claiming that he had seen eruptions on the moon. Then, according to Melosh, "In 1829 Gruithuisen proposed that the moon's craters had been produced by cosmic bombardment in past ages." That idea may have been taken more seriously if Gruithuisen had not damaged his credibility by also announcing the discovery of men, animals, and a ruined city on the moon, although none of this will come as a surprise to tabloid journalists that run amok in the United States in the late twentieth century.

Another red herring was dragged across the trail leading to a correct explanation of lunar craters by G. K. Gilbert, who in 1893 concluded that impacts were involved, but with a proviso. According to Melosh, "Gilbert performed experiments on low velocity impacts of various powders and slurries in his New York hotel room during the winter of 1891 when he was lecturing at Columbia University." He became convinced that the impacts had to be vertical in order to make circular craters. Because lunar craters are virtually all circular, and vertical impacts were not to be expected as a rule, the theory was dropped.

Elongated craters are produced in experiments with low-velocity projectiles that impact obliquely but when high-velocity objects strike the moon or the earth, the impact results in an explosion that produces a circular crater. This was appreciated by Ernst Öpik in 1916 in a paper "published in an obscure Estonian journal, written in Russian with a French summary." That meant it was missed by astronomers everywhere.

The scientific study of craters only took root after World War II when it was realized that impact craters would indeed be circular because of their explosive nature. Large asteroids don't just fall into the ground and push matter aside. Because of the rapid heating they experience while traveling through the atmosphere, they explode. Depending on their size the explosion may occur in the atmosphere, which produces a bolide, or at impact, which then forms a crater. Only in situations where the impactor arrives at a very low angle would an elongated crater be formed.

Figure 12-1 A 120-kilometer diameter impact crater on Mercury. Like all the terrestrial planets, Mercury has suffered a great deal of bombardment since its formation and there is no erosion on that planet to remove signs of past collisions. (Courtesy NASA)

Impact craters now turn out to be the most prominent geologic feature in the solar system. They are seen on Mercury (Figure 12-1), which is pockmarked by a stunning number, and Venus, under its dense cloud cover which prevents all but the larger objects from slamming onto its surface, as well as on the earth and moon, and Mars (Figure 12-2) and its two small moons (Figure 5-2). They are also seen on asteroids that have been mapped or photographed to date (e.g., Figure 3-5) and on the moons of the giant planets Jupiter, Saturn, Uranus, and Neptune. If nothing else, these pictures convince even the most skeptical that impacts have helped shape the surface of every solid object in the solar system.

The moon is covered in impact craters (Figures 12-3, 12-4) and over 300,000 with a diameter larger than one kilometer have been cataloged. Bearing in mind that the earth is considerably larger than the moon, the earth should have suffered over 8 million collisions capable of making Barringer-sized craters or greater. So where are they? Most of them have long since eroded away, or have been subducted by continental drift, which refers to one continental plate sliding beneath another to obliterate any geographical features that existed before.

In recent years, many impact scars have been found on earth. Richard Grieve of the Geological Survey of Canada in Ottawa has been keeping track of structures that are widely accepted to be impact craters. Some 145 craters, many of them found by geophysical study related to oil and mineral exploration, were identified by 1995. About 30 of them are deeply buried, covered by postimpact sediments. No impact structures have been identified from satellite photos alone. Only one confirmed impact structure has been found in an ocean; it is the 45-kilometer-diameter, 50-million-year-old structure off Nova Scotia. Another is

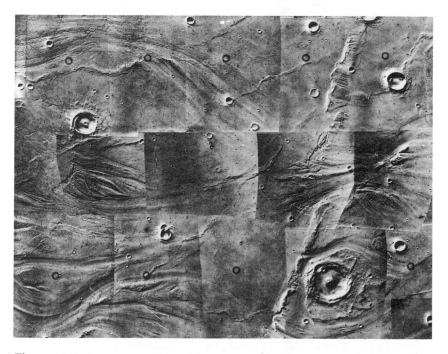

Figure 12-2 Impact craters on Mars scattered among patterns indicating that water once flowed over its surface. New work by Ann Vickery and Jay Melosh of the University of Arizona suggests that the Martian atmosphere was once thick enough to hold water and that may have been lost to space through comet and asteroid impacts that blew large volumes of atmosphere so far into space that the low Martian gravity could not draw it back. Impacts literally eroded away the atmosphere. (Courtesy NASA)

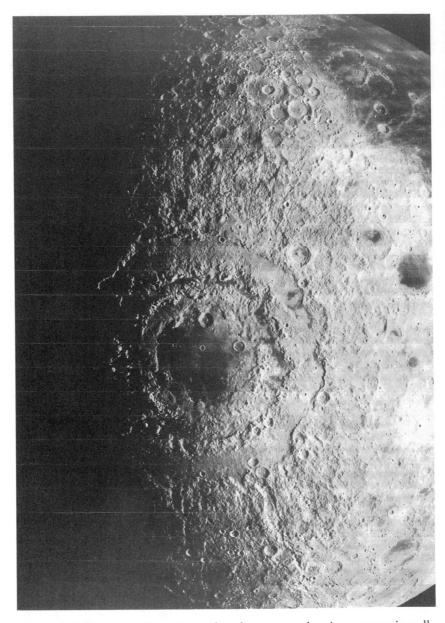

Figure 12-3 Bulls eye! Mare Orientale, a large crater showing concentric walls indicating an extraordinarily violent impact. The crater basin, which is located just at the moon's limb as seen from earth, has been filled with lava, which gives it a dark color. The near side of the moon has several such regions called maria that are so large that they can easily be seen with the naked eye. Even a pair of binoculars will help you see that the lunar surface is pockmarked by craters. The similarities between Mare Orientale and the Chicxulub crater (Figure 2-2) are striking. (Courtesy NASA)

Figure 12-4 The *Apollo 12* lunar module parked at the edge of a lunar crater on November 20, 1969. The *Surveyor III* spacecraft that landed there several years before sits among countless small craters formed by meteoroids that smashed into the moon unhampered by an atmosphere. (Courtesy NASA)

suspected in the Barent Sea. Figures 12-5, 12-6, and 12-7 show images of terrestrial impact craters.

Richard Grieve has said:

> It is apparent, in recent years, that impact can no longer be considered a process of interest only to the planetary community. It is a process that has affected terrestrial evolution in fundamental ways.
>
> Impact is the most catastrophic geologic process and is a natural consequence of the character of the solar system. It has happened on earth throughout geologic time and will happen again.

To make the impact crater list, the rocks in the crater should show evidence for great pressure (shock metamorphism). The planetary-wide distribution of these craters is shown in Figure 12-8.

Terrestrial craters are classified as simple or complex depending on their cross section. Simple craters up to a 2-kilometer diameter are produced if rocks

Figure 12-5 Terrestrial impact crater New Quebec, a 1.4-million-year-old, 3.4-kilometer diameter structure in Canada. (Courtesy Richard Grieve, Geological Survey of Canada)

Figure 12-6 The Gosses Bluff impact crater in Australia, a 143-million-year-old, 22-kilometer crater that has been severely eroded. The ring of mountains is what is left of the central uplift. The original crater was at least three times wider. (Courtesy Richard Grieve, Geological Survey of Canada)

Figure 12-7 The Brent impact crater in Canada, 450 million years old, diameter 3.8 kilometers. It is heavily eroded. (Courtesy Richard Grieve, Geological Survey of Canada)

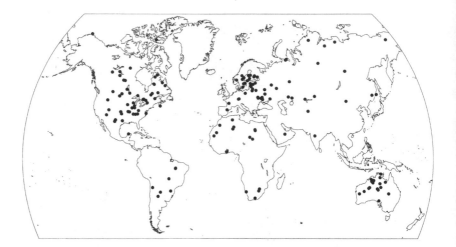

Figure 12-8 The global distribution of impact craters, clearly found all over the planet. The absence of any craters in the oceans reflects not only a lack of sufficient information, but also on the fact that craters produced in the ocean would likely be very difficult to find, even if they maintained their shape beneath thick layers of sediment that would tend to cover them in time. (Courtesy Richard Grieve, Geological Survey of Canada)

are sedimentary, up to a 4-kilometer diameter if rocks are crystalline. Complex craters have larger diameters. Simple craters involve a bowl-shaped depression with an uplifted or overturned rim area, such as Barringer. Crater diameters are typically 20 times larger than the size of the impactor. The depth of such craters, which may be filled with debris, is usually about one-third the diameter. It takes about 10 seconds to form a simple crater whose walls and floor are lined with melted and shocked material from the sides.

Complex craters have an uplifted center peak or ring. In some, the surrounding depression is partially filled with what was once molten rock material. This layer is 250 meters deep at Manicouagan in Canada. The final form of these craters depends on the strength of the rocks that were impacted. Also, the cavities produced by the impact tend to be filled in by the central peak collapse and the central peak itself is created by a rebound. Complex craters are therefore not as deep as simple craters relative to their diameter.

Terrestrial impact craters are currently being found at the rate of about 3 to 5 per year, with most in the past 20 years having been identified in Australia and the former Soviet Union.

There are many circular structures found around the world which various amateur geologists would like to see added to the list. The Carolina Bays of the south Atlantic coast of the United States are a favorite topic for enthusiastic amateurs. These are elongated depressions first identified in 1933 and immediately attributed to meteorites. According to Melosh, who has received more than his fair share of letters from proselytizers for this or that structure, their cause is no mystery. They represent phenomena related to permafrost, which is common in many parts of the world. What I find amazing is how much attention has been foisted on these bays by the community of interested lay people.

There is an alleged crater at Kilmichael in northern Mississippi. Its chief proponent, Mark Butler, a retired oil geologist, has made it his personal task to convince crater experts to take a closer look but so far they are not convinced. For those of us nonexperts who drive through the area, the evidence is unconvincing, but then surface appearances mean little in spotting impact craters larger than a few kilometers across. When one visits the Wells Creek crater in northwest Tennessee it is difficult to realize you are inside it unless you happen to know it is there and have a good geological map to show just where it is located. Wells Creek has a well-defined central uplift that is itself craterlike, about 5 kilometers across compared with about 13 kilometers for the outer rim where the ancient upheaval has left a dramatic ring of cliffs best seen up close.

Richard Grieve, who has also been subject of many a lobbying campaign to add a favorite near-circular structure to his list, will only accept one as an impact crater if there is evidence from rock samples that the area suffered an intense shock.

The impact crater at Manicouaghan in Canada is 214 million years old. The present crater is 65 kilometers across and encloses a circular reservoir around the central uplift. This crater represents the eroded remains of the original that was 100 kilometers across. According to Grieve, the crater was dug by an event that produced 10 to 100 times as much energy as flows out of the earth in a year in all volcanoes, continental drift, and earthquakes, yet there is no clear association with a mass-extinction event. To produce a crater this large, an object of at least 5 kilometers across must have been involved.

Grieve expects a cosmic collision capable of producing an impact winter every 100,000 years. Most species won't suffer, he says, but civilization's infrastructure will collapse. As British Prime Minister Margaret Thatcher said during a recent war in which Britain was involved, "The veneer of civilization is very thin." The danger to civilization posed by an asteroid impact is more in the form of a trigger. If 25 percent of the world's population were wiped out in the physical aftermath, I, for one, have little doubt that people doing unto other people in the bitter struggle to survive will then take care of a large fraction of the other survivors. This will be when "civilized man" may drive itself back into another dark age.

In 1994 C. Whylie Poag of the U.S. Geological Survey reported that an 85-kilometer-diameter crater had been found beneath the Chesapeake Bay. Estimated at about 35 million years old, it may have been the event that scattered tektites over much of the southern and eastern United States and parts of the

Figure 12-9 A string of craters on the moon photographed on the far side of the moon by the *Apollo 11* astronauts in 1969. The chain is about 50 kilometers long and the largest crater is about 17 kilometers wide. (Courtesy NASA)

Caribbean. In late 1995 it was announced that shocked quartz had been found in that structure, thus confirming its impact nature. The newspaper headline announcing this continued a trend to misrepresent totally what happened: "Meteor formed lower Chesapeake." The estimated 1.6-kilometer-diameter object responsible for the crater that formed the southern part of Chesapeake Bay was no meteor (see chapter 3), the name given to pea-sized objects that slam into the planet by the thousands every day. And in a continuing misrepresentation in the media of the consequences of impacts of such a nature, the *Washington Post* quoted a scientist as noting that if it happened today "Washington would probably cease to exist." Apart from the obvious joke as to whether anyone would notice, a collision with a 1.6-kilometer-diameter asteroid would trigger a global catastrophe that would terminate civilization (see chapter 13).

Three of the largest impact structures identified are paired with smaller ones. The 24-kilometer Ries crater in Germany has a partner, Steinheim, 46 kilometers to the southwest. Both are 15 million years old. The Kamensk and Gusev craters in Russia are both 65 million years old, and the twin Clearwater Lakes in Canada in northern Quebec, east of Hudson's Bay, are 290 million years old.

About 10 percent of the terrestrial impact craters are now suspected as likely doublets. At a meeting I attended in April, 1995, the possibility that most asteroids were doubles was being mooted. This raises the possibility that cosmic collisions of significant magnitude may not be isolated events. There is some evidence to indicate that the K/T boundary clay carries the signature of more than one impactor as well.

Double craters and strings of craters have been identified on the moon (Figure 12-9) as well as on Jupiter's satellite, Callisto (Figure 12-10). This suggests that objects that slam into solar system bodies may sometimes fragment before they hit. This obviously happened to comet Shoemaker-Levy before it struck Jupiter in 1994 (chapter 14), a fact that links back to the ideas proposed by Clube and Napier and their colleagues who developed the idea that earth's envi-

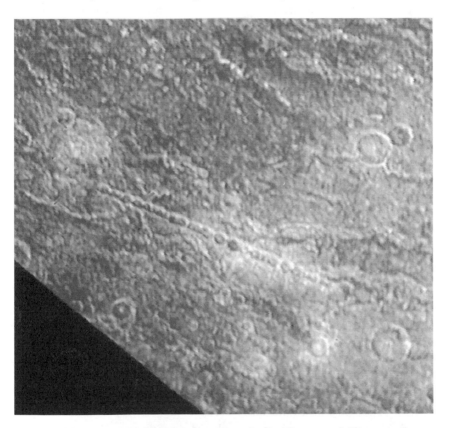

Figure 12-10 A string of impact craters on the Jovian moon Callisto stands out against the background of more random craters that mark its surface. Taken by the *Voyager* spacecraft, this image reveals one of about a dozen such chains on Callisto. This one is about 350 kilometers long and the largest individual crater is about 24 kilometers across. This image was downloaded from the World Wide Web for use in this book. (Courtesy Paul Schenk/Lunar and Planetary Institute)

ronment sometimes becomes hazardous because of the debris created when a giant comet breaks up after coming too close to the sun.

Two objects in close contact are not expected to separate enough just before impact to make a double crater, so why are double craters found? Paolo Farinella and his colleagues in Italy suggest that several close passes by the earth before impact would separate a contact binary so that the two objects would crash in distinctly different locations. Only fragments in close contact would be able to create crater chains like those shown in Figures 12-9 and 12-10.

There is another side to the terrestrial crater story; Grieve has collaborated with V. L. Masaitas of the Karpinsky Geological Institute in St. Petersburg, Russia to discuss their economic potential. "Impact is an extraordinary geologic process involving vast amounts of energy, resulting in near instantaneous rises in temperature and pressure, and in structural distribution of target materials. By the same token, economic deposits, in general, are the result of extraordinary geologic circumstances."

The Barringer crater was first explored in order to exploit its assumed iron-, nickel-, and platinum-rich core, and the same motivation drew Kulik's expeditions to the Tunguska region in 1929. In neither case was anything of worth found. However, of the more than 140 impact structures known, 35 have some form of potentially economic deposits and 17 of these are actually being mined for something other than building materials (Ries) or cement production (Ries), or exploited for hydroelectric power (Manicouagan).

Some impacts have helped bring to the surface important minerals, such as gold in the Vredefort structure in South Africa, uranium at Carswell in Canada and in Vredefort, and iron ore at Ternovka in the Ukraine. At Ternovka blocks of iron hundreds of meters across were fractured and tossed about by the impact and mixed with rocks that contain no iron. The Vredefort structure may originally have been 300 kilometers across and much of South Africa's gold wealth is associated with this 2-billion-year-old structure, whose riches set the scene for the growth of a nation.

In some cases the economic advantage is created because of the impact; for example, impact diamonds. They tend to be harder that normal diamonds and a variety called Carbonados has been sold for industrial uses.

Impact events sometimes alter the local geology to produce an economic resource; for example, oil is brought up from the 16-kilometer-diameter Ames crater in Oklahoma, which lies buried beneath 3 kilometers of sediments.

Drinking water is obtained from the Manson crater in Iowa and salt is mined in the Saltpan in South Africa. Oil is also found at the Red Wing Creek and the Newporte impact craters in the United States.

Over 50 percent of known terrestrial impact craters are less than 200 million years old. Older ones have tended to become lost in the continual upheavals that affect earth's surface.

What happens at impact? According to Jack Hills and M. P. Goda, if 22 kilometer per second is taken to be the typical impact velocity, then more than half the energy of impact of stony meteoroids with diameters less than 200 meters is absorbed in the atmosphere. The same is true of iron meteoroids of less than 80 meters in diameter. For a comet the diameter limit may be 1.6 kilometers. This causes these objects to explode or flare at the end of their visible paths. Most of the energy is dissipated in the atmosphere and no crater will result. The blast wave, however, will do considerable damage over wide regions.

Their modeling of the atmospheric effects shows that the plumes produced by asteroids larger than 120 meters across cannot be contained by the atmosphere so a bubble of hot gas forms above the atmosphere, which quickly girdles much of the planet. The so-called optical depth of the cloud, which is a measure of how much light is cut off, would be so great that no sunlight would penetrate to the ground. Even a 60-meter asteroid would produce enough heat radiation from the high optical depth, hot cloud, that it would ignite pine forests locally. The blast wave may then blow out the fire.

The most serious consequences to human life follow ocean impacts. An asteroid ranging in size from 200 meters to 1 kilometer slamming into the Atlantic would cause terrible destruction from tsunamis that wash over all its shoreline.

An ocean impact obviously produces a staggeringly large splash at the point of impact, even emptying the ocean to its bottom, but then enormous waves ripple outward. The height of the so-called deep-water waves does not change significantly as they radiate away, but when they enter shallow waters they crest and break as a tsunami. An average tsunami (Table 12-1) is 40 times higher than the deep-water wave that created it in the first place. Underwater explosion data have allowed Hills and Goda to check their computer calculations of what would happen if, for example, a small (200-meter) asteroid were to drop in the center of the Atlantic. Deep-water waves would be produced that would still be 5 meters high upon reaching shore. When they break, they will be 200 meters high. The wave pulse would last several minutes and would sweep over all low-lying land, including, for example, Holland, Denmark, Long Island, and Manhattan. Hundreds of millions of people would be wiped out in minutes.

TABLE 12–1 Height of deep-water wave
and (->) tsunami 1,000 kilometers away from point of impact

Size of impactor	*50 m*	*100 m*	*300 m*	*1 km*
Type of object				
Iron	2 m -> 80 m	7 m -> 280 m	40 m ->1.6 km	700 m -> 28 km(!)
Hard stone	0.8 m -> 32 m	2 m -> 80 m	25 m -> 1 km	200 m -> 8 km
Average time between impacts	100 yr	1,000 yr	20,000 yr	200,000 yr

Hills and Goda have offered an extraordinarily interesting personal speculation related to the past occurrence of tsunamis that may have caused widespread, and yet little-known, devastation along coastlines:

> These numbers are very disturbing to [us]. Perhaps the legendary tale of the lost civilization of Atlantis, which was said to be on the Atlantic coast and was engulfed suddenly by the ocean was due to such a tidal wave. It is somewhat surprising that there were no widespread coastal settlements along the Atlantic until after 800 A.D. when Vikings settled and fortified numerous towns along the Atlantic coast. The niche they exploited may have been opened by a previous disaster whose institutional memory has been lost.

It never made sense to me that Atlantis was supposed to have sunk beneath the ocean. It is much simpler to imagine that it was washed away.

The danger of being engulfed by an enormous tsunami is being recognized as more serious than anything that would be caused by a comparable land impact. As Hills and Leonard add: "Because a disproportionate fraction of human resources are close to the coasts, tsunamis are probably the most deadly manifestation of asteroid impacts apart from the very large K/T superkiller."

Their comments remind me of what the Tollmanns have suggested quite independently regarding the Deluge. They too have raised the possibility that Atlantis disappeared as the result of an impact tsunami. Both groups suggest more attention be paid to searching for evidence of ancient tsunamis along the coasts of major oceans.

By examining the Hills and Goda computer data we can figure out what would have happened if asteroid 1989 FC had struck the Atlantic Ocean 1000 kilometers from the coast of the United States. If it was an iron asteroid, the deep-water wave would have been 20 meters at the coast. The tsunami it would have produced would have been about 0.8 kilometer high. The notion of such a high wave breaking along the east coast of the United States is a stunning one.

While we are frightening ourselves with numbers, their calculations show that if a 10-kilometer iron asteroid, or even a comet about the size of the dinosaur killer, were to strike an ocean, the deep-water wave 1000 kilometers from the impact would be about 3 kilometers, which, in principle, translates into a tsunami over 100 kilometers high. With a deep-water wave that large, it is moot as to whether it would notice the shoreline and would, instead, wash over a good fraction of the continent and break up against the Rocky Mountains!

There are several other points that are important when we think about ocean impacts. The first concerns the likely creation of a hurricane of terrible scale. Kerry Emanuel of the Massachusetts Institute of Technology has made a model of what happens in the atmosphere after an ocean impact. The enormous heating of the water that bounces back from the splash creates so much of a disturbance in the water and the air above it that an enormous hurricane with winds of up to 800 kilometers per hour is produced. The eye of the storm might shrink to a mere 16 kilometers across. The swirling "hypercane" pumps a great deal of water up into the stratosphere, which later will fall to earth to produce horrendous flooding. This suggests more far-reaching repercussions of ocean impact than just tsunamis along coastlines. A "hypercane" makes the devastation even more staggering, although like normal hurricanes it would quickly loose energy as it moved over colder water, or over land, a relative concept that may be of little solace to anyone caught in its path.

13

OFFERING ODDS
ON IMPACT

*T*HERE is no doubt that the earth
continues to be struck by objects from
space. Most of the impactors are very
tiny, such as those that produce common meteor trails, and major collisions no
longer happen very often. But if a large object, a half kilometer across say, were
to strike our planet, the consequences would be devastating. In 1989 an asteroid
large enough to bring civilization to the brink of total destruction missed earth
by 6 hours and this close encounter in itself should be enough to give us food for
thought. There will be other close shaves in the years to come, but no one can
predict just when or how close. Only time will tell.

Fortunately there are several dedicated groups of astronomers around the
world searching for near-earth asteroids (NEAs) in order to catalog their existence and figure orbits lest any should be on a collision course. As a result of their
efforts, crucial data are being obtained that will allow the probability of impact
to be more accurately estimated, even if only in a statistical sense. The best anyone can do, or will ever be able to do, is to offer odds on the chance of collisions.

Odds on comet impact, in the form of estimates of the period between such
events, have been published for two centuries. Each generation no doubt felt that
the latest estimates were superior to those that went before. For example, in 1861
James Watson, in *A Popular Treatise on Comets*, said that "it has been found by
actual calculation, from the theory of probabilities, that if the nucleus of a comet

having a diameter equal to only one fourth part of that of the earth…the probability of receiving a shock from it, is only one in two hundred and eighty-one millions." This estimate was also quoted by Thomas Dick in 1840 who, in turn, credited it to Francois Arago for calculating this around 1800. To illustrate to his readers what this meant, Dick offered this analogy:

> Admitting then, for a moment, that the comets which may strike the earth with their nuclei would annihilate the whole human race, then the danger of death to each individual, resulting from the appearance of an *unknown* comet would be exactly equal to the risk he would run if in an urn there was only one single white ball of a total number of 281,000,000 balls, and that his condemnation to death would be the inevitable consequence of the white ball being produced at the first drawing.

In 1897 Herbert Howe had this delightful understatement in his textbook on astronomy: "While there are no definite data to reason from, it is believed that an encounter with the nucleus of one of the largest comets is not to be desired." In view of what is known about the threat today, we could not possibly disagree. He went on to estimate the chance of a collision and thought that the chance of an impact in the next 100,000 years was "exceedingly slight." On the other hand, "most astronomers would be delighted with the prospect that the earth was going to blaze a pathway through some ordinary comet." We cannot be sure what he meant by ordinary, but as long as the earth doesn't actually run into its nucleus and merely sweeps through the comet's coma or tail we might indeed have an opportunity to learn a great deal, without, at the same time, kissing good-bye to the planet.

Camille Flammarion in 1894 was less reticent about the probability of a collision. He said that it was unlikely before the natural death of the earth itself, but then a hefty bash from a large comet would surely spell that natural end (at least for life, which one supposes he meant), so the estimate is both correct and inadequate.

Specific odds were offered by Sir Richard Gregory in 1893: "about once in about twenty million years." In 1897 David Todd offered an estimate of an impact every 15 million years, but even if the earth should pass through the head of a comet, it might bring universal death to nearly all forms of animal life.

The famous textbook of Russell, Dugan, and Stewart in 1926 stated flatly that comet collisions would happen once in 80 million years. They imagined that if the head of a comet consisted of boulders weighing tons that the consequences would be serious, "although it would fall very far short of producing a wholesale destruction of terrestrial life." They were more concerned that the atmosphere would be poisoned if the earth ran into the outer portions of the comet. In his own book of 1935, Russell stated with equal assurance that there was no danger of an encounter between a comet and a planet. Decades later, in

1964, when George Abell's famous textbook (*Exploration of the Universe*) swept the market, we read that "the probability of collision between the earth and a comet is very small indeed." However, he admitted that in its billions of years of existence it may have occurred "even more than once."

In the past few years, the comet impact scenario has taken on a life of its own and the danger of asteroids has been added to the comet count. In the context of heightened interest in the threat, reassuring predictions have been offered about the likelihood of a civilization-destroying impact in the years to come. Without exception, the scientists who have recently offered odds have been careful in making any statement. They have acted in a "responsible" manner and left us with a feeling that the threat is not worth worrying about. This is not to criticize their earnest efforts, only to point out that estimates have been attempted for centuries.

The way I look at the business of offering odds is that it hardly matters whether the chance of being wiped out next century is 1 in 10,000, for example, or that the likelihood of a civilization-destroying impact is once in a million years. That's like betting on a horse race. The only thing that is certain is that a horse will win. What matters is the larger picture that begins to force itself into our imagination; comet or asteroid impacts are inevitable. The next one may not wipe us out in the coming century, or even in the century after that, but sooner or later it will happen. It could happen next year.

I think that what matters is how we react to this knowledge. That, in the long run, is what will make a difference to our planet and its inhabitants. It is not the impact itself that may be immediately relevant; it is how we react to the *idea* of an impact that may change the course of human history. I am afraid that we will deal with this potentially mind-expanding discovery in the way we deal with most issues that relate to matters of great consequence; we will ignore it until the crisis is upon us. The problem may be that the consequences of a comet catastrophe are so horrendous that it is easiest to confront it through denial. In the end, though, it may be this limitation of human nature that will determine our fate.

From this sobering perspective, let's take a look at the likelihood that our planet will be struck by a rogue asteroid or comet, and consider two groups of scientists offering odds who look at the issue very differently. The first group is predominantly located in the United States and has until recently described dangerous impacts as being caused by a lone wanderer through space that slams into the earth with little warning. I will refer to this as the Lone Ranger model, because it reminds me of Western mythology (that is, cowboy fiction) where all's well until the lone bad guy rides into town and begins to shoot up the citizens. In the movies, the town is rendered safe by a good guy, the Lone Ranger, doing brave deeds. In the case of future impacts, the brave deeds are being accomplished by those who are searching to identify the rogue asteroids and comets so that we might someday ride out and shoot them up (see chapter 16).

The other point of view is held by those astronomers who think that the greatest threat is posed by an unruly band of comets or asteroids that attacks en masse, like a Mongol horde emerging out of Asia, and which between them will produce as much damage to civilization as the giant can do in a single impact. The core of this group includes the aforementioned Musketeers in England and Australia.

Representing the Lone Ranger point of view, Clark Chapman and David Morrison published their odds in *Nature* in January 1994. In "Impacts on the Earth by Asteroids and Comets: Assessing the Hazard," they concluded that the chance that a large (2-kilometer diameter) object will slam into the planet and terminate civilization during the next century is 1 in 10,000. To put this another way, such an impact is a virtual certainty in the next million years. Of course, it could happen either then or next year, or sometime in between. That's the uncomfortable thing about these odds; they don't tell you when the winner—or in this case the loser—will come in.

Since their odds were offered, there has been a lot of debate about whether it would take a 2-kilometer object to plunge us back into a dark age, or whether a smaller one would do, something like a half-kilometer object. Based on what was seen after the comet impacts on Jupiter in 1994, it now seems fairly certain that a half-kilometer object would do nicely. That immediately ups the ante to something like once every 100,000 years.

If you like thinking about odds, about the likelihood of your winning a gamble, consider these statistics assembled by Chapman and Morrison from a large number of official sources. Their impact data translate the chance that a rogue asteroid or comet will spell any one person's death as 1 in 20,000, based on an average lifetime of 50 years. It turns out that the likelihood that any individual in the United States will die in a passenger plane crash is about the same. We spend a lot of money making planes safer to reduce these odds, but we do little or nothing to avoid impacts. Isn't that odd? At the very least we should follow the example of airport security and place metal detectors in earth orbit to prevent rogue objects from arriving unexpectedly, which is what astronomers searching for unwelcome visitors are doing in their own way. It so happens that their metal detectors (telescopes) operate over a distance, but we remain lamentably ill-prepared to disarm the intruders should the alarm be sounded.

The probability that someone in the United States will die in a flood is about 1 in 30,000 and the odds that someone will carve an epitaph on your tombstone that reads "Blown away by a tornado" is 1 in 60,000. This is three times less likely than your ending up at the receiving end of an asteroid, after which there will be few people left to carve epitaphs on anything.

Calculating the likelihood of a significant asteroid or comet impact is an interesting exercise of little practical value to you as an individual, unless you happen to be an insurance agent. A small impact on the east coast of the United

States triggering a few tens of millions of deaths and widespread destruction would probably drive most insurance companies into bankruptcy. When I asked my insurance agent to look into adding asteroid and comet coverage for our homeowner's policy, he stared at me blankly. After I explained, he promised to look into it. I never heard from him again. (I was probably lucky that he went ahead and insured me at all!) Upon closer examination, I found that the policy does not cover damage from waves or tidal water or overflow of a body of water. Even those of us living far inland would not be safe from a tsunami created by an ocean impact, after which it would become an interesting point of law as to whether a tsunami is a wave or can be regarded as an overflow of a body of water. But then again, if an impact sufficiently violent to trigger a tsunami was to wash away our house 800 kilometers from the coast, we might have a difficult time finding the remains of an insurance company's headquarters in any case.

When you think about it, a significant impact would wipe out so many cities and people and businesses that the concept of insurance becomes moot. What would they pay with? Who would get the benefits when society is plunged into barbarism because of the frantic struggle to survive by those who lived through the immediate aftermath of a significant impact?

A word of advice to insurance companies, however. I have heard it said that it is they who have the most to lose as the result of a small impact, one that takes out a few cities and their surroundings and leaves the rest of the country unscathed, something like a Tunguska explosion over New York, for example. The insurance giants may wish to invest in asteroid search programs for their own good.

To return to the statistics offered by Chapman and Morrison, the chance that you will die in a motor vehicle accident in the United States, counting a typical life span of 50 years, is about 1 in 100, so drive carefully. But like any of these statistics, that tells you little. It doesn't give you of a clue as to when it's likely to happen to you. You only realize that you are about to become part of the statistics just as the oncoming automobile hits you.

That is what it will probably be like for the human race. We'll appreciate the hard work put in by many scientists who tried to calculate the odds at about the time the thing from space looms above the horizon and begins to glow just before slamming into the earth.

Knowing your odds may help provide mental comfort, especially when you learn that you have a chance of 1 in 5000 that you will be electrocuted in a 50-year time span, not counting death by electric chair. That means in any given year the chances are about 1 in 250,000, comfortably safe, provided you're careful with electricity, and leave it to the thoughtless ones to make up the statistics.

When you really get down to it, we spend a lot of money in preventing all sorts of accidents, because they cost money as well as lives. We spend a great deal of time and money learning to forecast floods so that people will not be swept away, and the fuss we make trying to forecast tornadoes and creating a tornado alert system to protect ourselves is considerable. A few hundred yards from where we live there is a siren that sounds the alert when a tornado is imminent. We have no such alert to warn us of impacts. We even worry about possible death from fireworks (1 in 1,000,000), so why don't we get stirred up about an asteroid or comet cataclysm? One of those will do a lot more damage than take out a few careless individuals, or unlucky ones who were on the wrong airplane, near the wrong river, or living in an unfortunate trailer park in the path of a tornado.

Our perception of the threat of comets and asteroids suffers from what scientists call a selection effect. If the earth had suffered a significant regional impact that wiped out thousands of people in the past few thousand years, say before the nineteenth century, who would know about it except those who were killed? For most of recorded history the world has been so sparsely populated that even if thousands died almost anywhere in the world, other than in populous Europe, India, or China, from a Tunguska-like event say, who would have been be left to carry word to other parts of the planet? And if any survivors did stumble into another country to report the disaster, who would have believed their ravings about fire from the sky? Wouldn't such stories later be viewed with haughty disdain as the ravings of primitive people to which we could attach no value?

Over the last few years the Lone Ranger school may have underplayed the potential danger of asteroid impact. This became obvious when one of their number, a well-known asteroid hunter, when interviewed during the comet SL9 impact adventure of July 1994 was able to blithely state that if one of those fragments had hit earth it would have made a hole the size of Long Island. He did not add, and hence the media did not report, the subtle added distinction that if an asteroid was capable of producing a crater that size it would have produced a blast that would have terminated civilization.

It is one thing to figure odds of impact, but how can anyone estimate how many people will die in the aftermath? It turns out that some scientists have been thinking about large-scale nuclear war for decades and they have done a lot of calculating related to the consequences of such a catastrophe. Their computer simulations were meant to predict the after-effects of a worldwide nuclear holocaust. After the Alvarez team published their hypothesis relating to species extinction triggered by a cosmic impact, the nuclear experts realized they had to take into account an additional hazard: the consequence of injecting a vast amount of dust and smoke into the atmosphere, enough to darken the globe and cool the earth. That gave rise to studies of the so-called nuclear winter phenomenon and it soon became apparent that all-out nuclear warfare would do far more than destroy the enemy; it would destroy civilization.

Nuclear winter results when the smoke, soot, and dust raised by the explosions girdle the earth and plunge the planet into a long winter which, in turn, will cause more deaths than the initial blasts. It didn't take certain alert astronomers long to realize that the injection of energy from an asteroid impact would produce comparable results, an impact winter, without the nasty side effect of radiation sickness.

Awareness of the chilling parallels between nuclear holocaust and asteroid impact scenarios has lead to the threat of comets and asteroids being taken more seriously. The point is that the calculations and computer modeling have been done as regards what happens to our planet when you explode a certain tonnage of explosives, say 50 or 1000 megatons, or even millions, in the atmosphere. It is also possible to calculate the energy released by an object of a certain mass traveling at a high speed when it hits the earth. For example, an object 1 kilometer across traveling at about 40,000 kilometers per hour would produce a 15,000-megaton explosion, or thereabouts, depending on the density of the object. The 10-kilometer-diameter dinosaur killer produced a blast of about 100 million megatons of TNT. Knowing these energies, it is possible to figure out how many fatalities will result from the impact of different-sized objects slamming into earth. There is, unfortunately, only one way to test the accuracy of such predictions.

In July 1994 an interesting article appeared in *Scientific American* on what was learned from the *Apollo* moon landings. The author, G. Jeffrey Taylor, a geophysicist at the University of Hawaii in Honolulu, pointed out that for every impact crater you see on the moon there would be 20 expected on the earth, because of the earth's larger size. He added that Frederick Hörz of the Johnson Space Center estimates that there are 5000 craters on the moon larger than 5 kilometers across which were produced in the past 600 million years. Putting these numbers together implies that half-kilometer objects, which are potentially civilization-destroying, are expected on average every 6,000 years. This rate is consistent with the claims made by the Tollmanns (chapter 8). If earth is hit by a 0.5-kilometer object every 6,000 years, roughly four out of five would be expected to land in the oceans, which means it is very likely that a major flood event would have occurred at least once in the past 10,000 years.

Duncan Steel has given a cogent summary that illustrates what is at issue when figuring impact odds. The following is his description of what may have happened in recent history, which is what will determine what happens in the future.

The upper size limit for cometary bodies may be debated at length, but the observation during recorded history of several comets at least 100 kilometers in diameter (e.g., Comet Sarabat in 1729) should have made it clear by the 1970's that giant comets exist, thus implying that the majority of the mass delivered into earth-crossing orbits is held in such objects. Evidence accumulated in the past two decades that confirms this view includes the discovery of a profusion of small comets in the Kreutz group (implying a massive progenitor which broke up presumably due to tidal forces or thermal stress when in a sun-grazing orbit), the

100–400 kilometer asteroidal/cometary objects 2160 Chiron, 5145 Pholus and 1993 HA2 in orbits crossing the outer planets (such orbits being liable to diversion into earth-crossing paths on time-scales of approximately a million years), and, to the time of writing, eight 100–400 kilometer objects in a reservoir just beyond the planetary region.

Here he is referring to the objects in the Kuiper belt. This raises the issue of where the short-period comets and near-earth asteroids come from. Steel puts it this way:

If a giant comet had arrived in the inner solar system within the last 100,000 years and undergone a hierarchical disintegration similar to that of comet SL9 but on a trans-earth/Jupiter orbit, having implications for climatic deviations and our recent and continuing impact history, then it could have produced not only the entire short-period comet population, but also a major fraction of the near-earth asteroids.

He adds that at least one giant comet arrived in the past 20,000 years and that its remnants are still with us, in the meteoroid streams that are part of the Taurid Complex, in particular the dominant Taurid Stream itself, and adds that "Any investigation of the ecology of near-earth objects and their implications for the contemporary hazard faced by mankind which ignores this evidence is therefore based upon an important, but apparently erroneous, assumption."

A Tunguska type event may be what is needed for us to take seriously this threat from space. A small impact capable of wiping out a country such as Paraguay or a state such as Virginia would, in this era of mass communication, force us to take notice. With millions killed, the resources of neighboring states or nations would be pushed to their limits in coping with the tragedy, whose consequences would no doubt escalate because of the sheer scale of the devastation that would be produced. Then imagine the reprisals and the questions about why we didn't do something to prevent such a catastrophe.

I have already referred to Chapman and Morrison's estimate of the odds of a major impact next century, a probability something like 1 in 10,000. But there is plenty of room for making better estimates as the data base improves. For example, a hint that others are beginning to consider the role of tsunamis as significant recently emerged in a report of the Planetary Defense Workshop held at Los Alamos in May 1995. David Morrison posted a summary on the World Wide Web which made it clear that Edward Teller and colleagues were paying increased attention to the work of Hills and Goda, referred to in the previous chapter, and that the likelihood of a devastating tsunami created by an ocean impact might be as low as a few percent in the next century. This converts to one tsunami catastrophe every 5,000 years, consistent with our estimate in chapter 12.

Duncan Steel suggests that the comet impacts on Jupiter (chapter 14) may even have something to offer as regards calculating odds. Jupiter is likely to be struck about 1000 times as often as the earth and there is tantalizing evidence for dark markings such as followed the July 1994 impacts having been seen on at least four previous occasions: in 1885, 1928, 1939, and 1948, or five collision events per century. That would imply that earth might be hit every 20,000 years.

While researching old books and technical articles for this chapter, I noticed something fascinating that illustrates how our conception of what the future holds depends so intimately on our knowledge base. In two centuries, the typical estimated time between comet impacts (from old books) to impacts capable of producing global catastrophe (from new research) has decreased from once every 281 million years (which held for most of the 19th century) to about once ever 5 to 10,000 years in the past year.

Compared to estimates made in the past century, one thing has changed in recent years. The NEAs have entered the picture and therefore the odds of impact have shortened dramatically. Around the end of the nineteenth century the odds had shortened to once every 10 million years or so, which held until the early 1980s when 100,000 years between civilization-destroying impacts began to surface. That change happened because of the sudden increase in information about NEAs, crater statistics, and past mass-extinction events.

A spate of at least nine estimates appeared in early 1995, four of which independently set the interval between such collisions at close to 5,000 years. If we extrapolate the recent trend toward a rapid decline in this number, it will not be long before we had better duck! This is not to suggest that the trend will continue, because there is little room for that to happen. After all, a significant impact every century, something very much greater than Tunguska, doesn't seem credible, even if it cannot be totally ruled out on the basis of more recent historical data, which continue to be lamentably scarce as regards the impact phenomenon. Perhaps we are, after all, in a lull between major catastrophes, as implied at by the Musketeers in their study of the Taurid Stream.

If major impacts capable of threatening civilization on a global scale do indeed occur on average every 5,000 years, this implies that at least two should have happened since the end of the last ice age. Even if we accept the Tollmann hypothesis of a major event 9,500 years ago, that still leaves room for one or two since then. This also ties in to the suggestion made by Clube and Napier that impacts affected the evolution of civilization about 5,000 years ago. In addition, we may ask whether significant impacts, especially those capable of triggering large tsunamis, would even have been noticed 500 to 1,000 years ago, except by those who perished, especially if a local disaster took place in a far corner of the globe. Seeking answers to these questions may yet offer a fruitful area for study

I agree with Steel when he writes that "simple probabilistic arguments cannot be used in our present state of ignorance, there being so many things we simply do not understand." At the same time we find little solace in knowing that

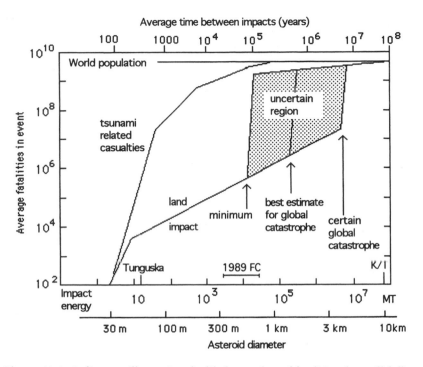

Figure 13-1 A diagram illustrating the likely number of fatalities that will follow impact by asteroids of a certain size, based on a similar plot given by Chapman and Morrison and modified by the author. The consequence of ocean impact producing large tsunamis is indicated by the upper curve, which has been based on the computer simulations of wave size from Hills and Goda and limited data of actual fatalities produced by historical tsunamis, which were very localized (see text). The estimated sizes of the K/T and Tunguska impactors as well as asteroid 1989 FC, which missed earth by 6 hours, are indicated. The shaded area indicates the highly uncertain regime where global catastrophe, as defined by civilization's ability to survive the impact.

scientists are inherently conservative creatures when it comes to making predictions. Unfortunately, the only way we will learn how accurate any of the estimates are is to await the next flurry of activity from the heavens.

To conclude this chapter, I want to show how the ocean wave model calculations of Hills and Goda (chapter 12) modify the conclusions of Chapman and Morrison concerning the likelihood of our being wiped out in a future impact. In Figure 13-1 I have redrawn a diagram given by Chapman and Morrison that estimates the number of casualties following impacts by asteroids of various diameters.

The fatalities created by a water impact depend on where the object crashes relative to land. If it happens close to shore, the damage along a limited stretch of coast would be staggering, whereas a midocean impact would produce a slightly smaller tsunami acting over a far larger circumference of the entire ocean.

In Table 12-1, the Hills and Goda calculations for the size of the deep-water wave produced at a distance of 1,000 kilometers from the impact by either an iron or a hard stone asteroid were summarized together with the height of the tsunami breaking on the shore (40 times larger than the deep-water wave, according to them).

Data on tsunami fatalities are, fortunately, sparse. However, we do have some to go by; in 1896 a 24-meter tsunami killed 22,000 in Honshu and in 1883 the Krakatoa tsunami 35 meters high killed 36,000. These were highly localized waves. If they were to girdle the Pacific as the result of an impact, tens to hundreds of millions would be drowned around its shores.

Tsunamis in the 20- to 35-meter range are produced by deep-water waves in the 1- to 2-meter range, and they, in turn, by an impact of an object in the 50- to 100-meter-size range, comparable to the object that created the Barringer crater or the Tunguska event.

The tsunami height drops very rapidly below this size range, which suggests a critical impactor size above which the danger suddenly becomes very great. That is why the additional tsunami curve drawn in Figure 13-1 is so steep.

The chances of being struck by an object 100 meters across is about once in 1,000 years, based on crater data as well as estimates of the population of NEAs in this size range, and it is now estimated that there are more than 150,000 objects like this in earth-crossing orbits. The number of casualties resulting from an ocean impact of such an object may be 10,000 times larger than given by Chapman and Morrison (Figure 13-1). No matter how you look at it, the danger posed to human life is clearly greatest from such an ocean collision.

Ocean impacts occur three to four times more frequently than on land. This is why I think flood legends should be the norm around the world rather than the exception. This is why I give credence to Clube and Napier, as well as to the Tollmanns, even if details of their hypotheses may be subject to argument.

To wipe out a quarter of the world's population, the point at which civilization's survival is surely threatened, an impact in the north Pacific producing a kilometer-high tsunami might do the trick. That requires an asteroid in a size range of a few hundred meters. Something like that is likely to smash into our planet every 10,000 years, on average. The chance that this will happen next century is therefore 1 in 100, which is 100 times more likely than the Chapman and Morrison estimate based on assuming a land impact of a 2-kilometer object.

This estimate is consistent with one reported by Duncan Steel. He says that "Shin Yabushita of the University of Kyoto, who has good reason for interest in the topic since Japan is especially at risk, has calculated that there is at least a 1% chance that practically all the cities around the Pacific rim will be obliterated within the next century by an asteroid-induced tsunami."

These estimates lead to an inevitable conclusion that has barely been discussed by biologists, at least not yet. Tsunami events must have played a very important role in directing the course of evolution of *Homo sapiens* over the past four million years. Many branches in the chain of human ancestry may have been wiped out by planetwide floods during that time. During the four million years since human ancestors roamed the planet, about 20 objects in the 1-kilometer-size range should have slammed into the earth, of which 6 would be expected to have crashed onto land and 14 in the oceans.

Land impacts would have made craters about 10 to 20 kilometers across. In fact, four in this size range have been found. They are at El'gygytgyn in Russia (18 kilometers; 3.5 million years old), Bosumtwi in Ghana (10.5 kilometers; 1.03 million years old), Zhamanshin in Kazakhstan (13.5 kilometers; 0.9 million years old), and Aorounga in Chad (12.6 kilometers; 4,000 years old). Given the apparent incompleteness of crater data, this is surprising agreement. The implication is that the role of impact events in hominid evolution should not be ignored and that our ancestors somehow managed to survive major flood events, which may explain why their remains are found in the depths of Africa, for example, rather than along its coast lines. Tsunamis would have acted as giant brooms keeping coastlines clean.

The role of smaller ocean impacts is also important. They may have done less damage to coastlines, but would have wiped out many communities that lived there. Until a few centuries ago, the number of people living along the shorelines of most of the world's oceans was relatively low and two hundred years ago an impact-generated tsunami in a remote coastline like that of South America or Africa, might have cost relatively few lives. The disaster might even have gone unnoticed, except by the victims, which brings us back to the ideas suggested by Clube and Napier, and the Tollmanns.

When we put the statistics of impact probability summarized by Chapman and Morrison together with the computer simulations of ocean impacts by Hills and Goda, we cannot avoid the conclusion that massive flooding must have been experienced in various parts of the world several times during the last 10,000 years. It is meaningless to suggest we are overdue for an impact; the next one could be here any day, or arrive a few thousand years hence. Nevertheless, it would be interesting to seek evidence for such events in more recent times than have been considered by either Clube and Napier, or the Tollmanns.

I, for one, will not be surprised if within a century a very large number of people die as the result of an asteroid smashing into an ocean. If they have enough warning, some people might be able to escape the worst of a wave heading toward their shores. This bears on a point I will make again in chapter 17: when an asteroid threatens, you either push it aside or you get out of the way.

The key issue about comet or asteroid impact is not when will it happen again, but given that it has happened and will happen again, what does this awareness tell us about our place in space? How does this affect our awareness of our existence in a larger context, in the cosmic scene? Or does it make no difference to us to discover, beyond a shadow of doubt, that we came into existence in a violent universe and that we will leave by the same route, collectively?

I predict that as we enter the twenty-first century astronomers will pay far more attention to what is going on close to home than they now do. The focus of research may yet swing from the current obsession about what is going on at the edge of the universe to what's going on here, because we will increasingly have to confront the fact that random fly-bys of comets and asteroids pose a real threat to our peace of mind.

I suspect that the earth has a good chance of experiencing another Tunguska event within the lifetime of most people reading this book. To put this another way, I will not be surprised if it happens. All I hope for is that, when the moment comes, I won't be caught with the phrase "I told you so" half out of my mouth before the blast hits. I'd like to be around to see how we cope with the catastrophe.

14

THE GREAT COMET
CRASH OF 1994

*J*UST as everyone offering odds was beginning to feel secure that comet impact is not an immediate threat to life on earth, I heard several planetary scientists state confidently that there was not even a chance of our seeing a comet-planet collision in our lifetimes. Since then Jupiter suffered the humiliation of being lashed by no less than 21 comet fragments, most of them sizable objects in their own right.

The saga began on March 24, 1993, when Carolyn Shoemaker discovered a "squashed comet" in a photograph that had just been taken by the team of which she was a member. She and her husband, Eugene, together with David Levy were using the 18-inch Schmidt telescope on Palomar Mountain near San Diego to hunt for near-earth asteroids and none of them had ever seen a squashed comet before.

What happened to set the scene for the discovery of the squashed comet is now apocryphal. The previous night had been perfect for the search, but someone had apparently exposed a box of film to daylight so that the exposures turned out totally black. The box of film was set aside, which was no immediate loss because on the fateful night the skies were mostly cloudy, which made comet hunting very difficult. Apparently David Levy, not wont to waste any opportunity to take more pictures, suggested that they go ahead and use some of the ruined sheets of film just in case some were not totally useless. At a cost of $4 per sheet, the film was usually very carefully used, given that they took photographs every ten minutes or so, all night long. Normally they would not have taken data that night, but why not go ahead and fire off a few exposures with the bad film. between and through the clouds.

As luck would have it, the sheets of film deeper into the box were only light-damaged around their edges, so they managed to get some nice photographs.

It was Carolyn Shoemaker's task to place a pair of developed pictures taken 45 minutes apart of the same area of sky into a stereo microscope. When viewed in this way the fixed stars produce no depth impression but an asteroid or comet moving with respect to the stars appears as if suspended in the field of view. And there it was—an image of something moving through space. She later said "It was terribly exciting. I had just about abandoned all hope of finding anything." Their reaction: "We were all sort of stunned."

The next step in a comet hunters' game plan is to report the discovery of a new comet to the International Astronomical Union's Central Bureau for Astronomical Telegrams in Cambridge, Massachusetts, where the keeper of the records, Brian Marsden, passes on the news to those astronomers around the world who want to be kept informed of comet discoveries.

Soon there was confirmation from Jim Scotti at the University of Arizona's Lunar and Planetary Laboratory, who pointed its 36-inch Spacewatch telescope at the new comet. He saw what would soon be trumpeted around the world; the squashed comet was a string of 21 fragments of what had surely been a single object at one time. Many comets have been known to break up after a close encounter with the sun, but they usually fragmented into two or three pieces. Twenty-one was unheard of.

Before we go on it must be said that there are many comet watchers and seekers in the world. If the Shoemaker-Levy team had not spotted this one, the honor would have gone to someone else. Brian Marsden later reported that within days independent reports of this comet had been received. But to the discoverers goes the honor of having their names attached to the new comet. So this one came to be called Comet Shoemaker-Levy 9, the ninth in their stable of discoveries. It will be referred to as comet SL9 in the rest of our story.

It takes several nights of observation to derive an orbit for a comet or an asteroid. A number of positions of the object have to be measured with respect to the background stars, and then those scientists who know how to manipulate the relevant mathematical equations in a computer are able to derive an orbit. Brian Marsden is an expert in calculating orbits and it didn't take him long to announce that comet SL9 was orbiting Jupiter, not the sun.

A few days later the orbit had been refined. Comet SL9 was in a two-year orbit around Jupiter and it had come so close in July 1992 that it had been broken up by the planet's gravitational pull. As Mike A'Hearn of the University of Maryland later became fond of pointing out, comet SL9 was not really a comet at all, at least not since it entered into orbit about Jupiter. From that moment on it was a satellite of the giant planet, a moon with a very interesting orbit.

Orbit calculations showed that in July 1994 SL9 would try to pass within 50,000 kilometers of Jupiter's center but, because its radius is 71,900 kilometers, this meant the comet was going to try to fly through the planet. That would produce what in the trade is known as a collision. Suddenly astronomers were confronted with the opportunity to observe what had become the subject of so much debate since the discovery of the iridium layer associated with the extinction of the dinosaurs.

The bandwagon began to roll immediately. Predictions of impact times soon spread around the world, into the newspapers and news broadcasts on radio and television. Before long, estimates were made of the size of the object that broke up to spawn the 21 daughters. It must have been about 10 kilometers across, some astronomers said, which, for those who carried such numbers in their heads, sounded ominously like the size of the object that struck earth 65 million years ago. The individual pieces were estimated to be from 1 to 4 kilometers across and their impact would produce some almighty explosions. Others, however, argued that the parent had been smaller, more like a few kilometers across.

The collision of SL9 with Jupiter would give just the sort of data that would allow the computer simulations of asteroid and comet impacts on earth to be tested. Although SL9 was going to strike a gaseous planet with no land masses, that didn't matter. It is what happens in the atmosphere that is most important, even on a planet like earth. And by remarkable good fortune, this spectacle was going to be witnessed from a safe distance of a billion kilometers.

All this also meant that comet SL9 would cease to exist a little over a year after being discovered. Carolyn Shoemaker said that she was not used to losing comets. They usually paid their respects to the sun and then to left for the depths of space, relatively intact. In confronting the potential loss, she expressed the hope that "it will be a grand show. If I'm going to loose a comet, then I want it to go out with fireworks." That is just what it did.

The demise of comet SL9 was observed by more telescopes, worldwide, than had ever been trained onto one place in the sky at any one time. The consequences were seen by more amateur astronomers and lay people than any other single astronomical phenomenon ever observed. Above all, even those who had no access to a telescope suitable for viewing an impact managed to enjoy the thrill almost as soon as the observers themselves, thanks to the information superhighway known as Internet.

At the University of Maryland, Mike A'Hearn set up a marvelous network of e-mail links to all those who planned to use their large telescopes to observe the impacts. Because those observers quickly shared what they saw, many of us heard from the front lines almost immediately.

The impact sites were predicted to be just around the planet's edge and would rotate into view within 20 minutes of each collision. The best we could hope for was to see some sort of aftereffect when the impact site rotated into view.

As the fateful day approached, predictions flew left and right. If the object that had broken up was a comet, the fragments might be fluffy and might fizzle on impact, like marshmallows slamming into a bucket of water. On the other hand, if the fragments were made of solid metal or rock, the individual explosions might be gloriously violent. It took courage to make any sort of prediction at all. Paul Weissman at the Jet Propulsion Laboratory in Pasadena was the only one to go out on a limb to predict something no one wanted to hear. In the British scientific journal, *Nature*, he stated flatly that "the Big Fizzle is coming." He had calculated what sort of cometlike object could have broken into 21 fragments and concluded that it had to have been very clumpy to start with, which meant that the individual fragments headed for Jupiter were equally clumpy. Instead of being hit by a bullet, Jupiter might be hit by buckshot. It turned out that the limb he had climbed onto was unceremoniously sawed off as reports began to come in over the Internet.

Because I'd like to follow the saga as it developed over the Internet in July 1994, it is worth describing just what is possible with this electronic network. The Internet connects virtually every college, university, government agency, and national laboratory computer network with every other such network, all over the world. Increasing numbers of commercial users have tied in as well, so that there now exists a vast web on connections that allow one to contact colleagues virtually anywhere, and essentially instantaneously.

For the Jupiter encounter, students at the University of Arizona created a "node" on the World Wide Web (a network of connections within the Internet) to archive images contributed by observers of the comet SL9 impacts. A similar resource was set up at the Jet Propulsion Laboratory in California. Anyone with access could download images from these various sources, which is how I came to observe the SL9 impacts, live from outer space. It later turned out that over 3 million connections were made by people all over the world to the Jet Propulsion Laboratory node alone during the week of impacts to download images.

The enormous interest in the collisions was attested to in reports sent from the heart of China and from Hong Kong by Jeff Greenwald, one of the world's first Internet travelers. He was engaged in an around-the-world journey by land and sea carrying a laptop computer and a modem. From time to time he sent reports to a friend in New York who placed them on the World Wide Web. Greenwald happened to be in China just before the Jupiter impacts and sent this report:

July 21, 1994/Chengdu, Sichuan Province, P.R. China

There's strange comfort in knowing that, even halfway around the world, cosmic phenomena exert an irresistible grip on the human imagination. The front page of today's issue of the local Chengdu paper features a huge artist's perspective of Jupiter, seen from a comet's-eye view. I suppose that eyes all over the planet have been oriented in that general direction, as humanity enjoys its third collective sky-watching craze in 25 years (I'm counting the moon landing and non-event Halley as the other two). No one is immune. Last week, when I was camping out in the highlands outside of Songpan, the local horsemen borrowed my Nikon 9x25 binoculars and squinted raptly upward, shouting to each other in the local dialect as they resolved wobbly glimpses of the fifth planet. (Which despite the cataclysmic drama rippling its methane seas, couldn't compare with their eye-popping, mind-blowing first view of our very own Moon.)

(It should be pointed out that Jupiter does not have methane seas; it is a gaseous planet whose atmosphere consists mostly of hydrogen and helium with a trace of methane as well as ammonia thrown in for good measure.)

After it was all over, on August 1, I downloaded this from Jeff, who was in Hong Kong at the time:

> This month's movie—screening three times every afternoon in the theater of the Hong Kong Space Museum—is Comet Crash! The film details the discovery of comet Shoemaker-Levy, and stimulates, with wicked computer graphics, July's cataclysmic collision between the fragmented space debris and the giant planet Jupiter.
>
> I went to see the movie this afternoon, but it was completely sold out. I wasn't surprised. One couldn't pick a better metaphor for the deepest fears lurking in the subconscious for nearly every citizen of the Territory [as they prepare to be handed over to China in 1997].

It seems everyone, everywhere, heard about the impact between SL9 and Jupiter. This is how I saw it happen.

On the evening of July 16 I gave a public lecture in preparation for the crash. The audience was supposed to view Jupiter through a telescope, but that hope was doomed by cloudy weather. What we did get to see was almost as exciting. The organizers of the event had obtained a feed from the NASA Direct cable channel, which broadcast a remarkable news conference by the Shoemakers and Levy. Old-fashioned, exciting astronomy was being reported from NASA without hyping spacecraft or shuttles.

Eugene Shoemaker animatedly reported the early news from Spain and Chile. Infrared telescopes had seen a bright plume from impact A on the limb of Jupiter. The crash was no fizzle.

After I gave my talk, we returned to the television monitor and watched another press conference. What an experience it turned out to be. We learned that observers at several observatories had seen the aftermath of the impact of fragment A. It was clearly going to be one heck of a week. A BBC reporter asked what these events meant to the man in the street back here on earth. One astronomer alertly replied, "He can be thankful that he's not on Jupiter."

When I got home, I accessed the University of Maryland bulletin board and read: "FIRST POSITIVE REPORT received here at 16:29 EDT from Calor Alto observatory, in Spain."

Impact A was observed with the 3.5-m telescope at Calor Alto using the MAGIC camera. The plume appeared at about nominal position over the limb at around 20:18 UT. It was observed in 2.3 micron methane band filter brighter than Io [one of Jupiter's moons]. Signed: Tom Herbst, Doug Hamilton, Jose Ortiz, Hermann Boehnhardt, Karlheinz Mantel, Alex Fiedler.

Calor Alto made a name for itself on the Internet by sending out reports from the three telescopes it had in action. They observed for 10 hours a day for 7 days, usually in the infrared (micron) band, and every fireball they recorded was different from all the others. As Tom Herbst said in showing one of their images later, "You can see Jupiter pretty badly damaged at this point." When you stop to think about it, that is an awesome concept, a badly damaged Jupiter, given that it is 300 times more massive than the earth.

From La Palma island in the Atlantic Ocean came confirmation.

The impact site of fragment A is visible in images obtained with the 1-m Jacobus Kapteyn telescope on La Palma from 22:38 UT onwards. Related cloud disturbances are also possibly visible, including a suggestion of a partial ring-like structure (not agreed on by all observers!) expanding into surrounding belts. However if real, then the later feature appears to be centered at a lower latitude than the impact site itself.

Impact B, which was the one we had hoped to see in Memphis, turned out to be a bust. "Observations from WIRO [the University of Wyoming Infrared Observatory] show no obvious evidence of a plume from fragment B." (In the interests of brevity, names of the countless observers involved in this and the following reports have had to be edited out.)

From Chile: "Hello world. It's cloudy today at Las Campanas, but last night it was clear enough to obtain an interesting result. *Io appears to have changed color* during the time that fragment B impacted Jupiter. The spectrum became increasingly red starting around 2:50 UT (July 17 1994) compared to its color 20 minutes later and 3 minutes earlier (when we started the observations…just in time!)."

Then the big boys and girls got their say:

The Hubble Space Telescope Team reports a location for the feature created by the impact of fragment A = 21….At 410 nm [nanometers, a measure of wavelength] the maximum radial extent of the dark region surrounding the streak is 12,000 kilometers. Although the region appears to be spatially circularly symmetric, the intensity distribution appears to be asymmetric, being more apparent to the south. A bright feature appears detached 1000–1500 kilometers above the limb on July 16.846 UT in the 953-nm filter, but it is not present in an image 3-min. earlier. A possible interpretation is that the feature is visible by reflected sunlight and that the apparent detachment is due to the shadow of Jupiter on the plume.

At the Space Telescope Science Institute someone wisely videotaped the group waiting to see the first images. When the fireball came into the telescope's field of view and the image began to build up on the screen, wild cheering was tinged with a sense of relief. The SL9 impacts would not be a dud as some had predicted. It was going to be a spectacular show. Break out the champagne. Heidi Hammel had come prepared and the cork was popped. Then everyone searched for glasses. They were ready with the world's most expensive telescope to photograph Jupiter's humiliation beneath the onslaught of debris from space, had brought the champagne, but no one had brought glasses or cups. Undaunted, they swigged from the bottle. (See Figure 14-1 for a trio of Space Telescope images.)

Further net-surfing produced:

Fragment E (predicted at 1505 UT) CCD imaging at Perth Observatory July 17 1450–1530 UT. Broad CH_4 filter at 893-nm, 2-sec exposures of limb of Jupiter (rest of Jupiter occulted by mask) with 60–70% duty cycle. No flashes or plumes were obvious in the raw data. Flashes greater than 25% of a square arcsec of Jupiter should have been seen in the raw data. Mike A'Hearn.

This was sad. It sounded as if the organizer of the University of Maryland bulletin board had traveled to Australia and was not seeing the impacts live from outer space.

Impact D was reported from the Anglo-Australian Telescope. "This flash was about 4 times brighter than the fragment C impact site." And C had been brighter than A.

From South Africa:

Soon after the predicted time of impact of fragment E, a bright plume at the limb of Jupiter was seen at the SAAO [South African Astronomical Observatory] in the K band images. The development was similar to that of fragment A. As the impact site rotated into view a large bright spot, about half the diameter of the Great Red Spot was seen on the surface. Soon thereafter a second spot appeared at the limb and when it rotated further it was seen to be centered in a dark circular patch. By about 1800 UT a third spot appeared at the limb, again apparently surrounded by a dark patch. Meanwhile, a dark circular feature centered near spots 1 and 2 could be seen easily in the eyepiece of the 0.75 m telescope at SAAO, Sutherland. Stay tuned.

Then the results began to pour in from the South Pole where an infrared telescope, SPIREX, operated by the University of Chicago had the benefit of a long winter night to watch the impacts. The astronomers at the Pole made 1000 images per day and obtained essentially continuous data except for two snowstorms. Within 30 hours of their taking data at the South Pole of fragment A,

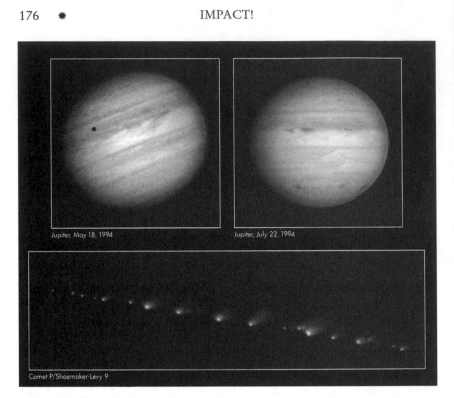

Jupiter, May 18, 1994

Jupiter, July 22, 1994

Comet P/Shoemaker-Levy 9

Figure 14-1 Comet Shoemaker-Levy 9 observed by the Hubble Space Telescope. Lower image, the comet train on May 17, 1994, showing the 21 fragments stretched along a million km, three times the earth-moon distance, heading for Jupiter. *Left.* Image of Jupiter taken on May 18, 1994 showing the "undisturbed" planet. The dark spot on the disk is the shadow of the inner moon, Io. *Right.* After the impacts on July 22, 1994. Dark smudges are seen beneath the famous red spot marking. The impact sites of 6 fragments can be identified here (A, E/F, H, Q1 and Q2). (Courtesy Hubble Space Telescope Team/NASA)

their photograph was on my desk, thanks to Internet and the link created at the University of Maryland.

> The South Pole Infrared Explorer (SPIREX), a 60 cm telescope, has detected the plume from fragment A at 2.36 microns. At 20:25 UT it was still brighter than Io. Brightness diminished rapidly during the next two minutes, but it remained visible at this wavelength for 20 minutes. Images at other wavelengths are being transferred back from the pole currently.

Later:

> SPIREX detected a fragment C impact site at 07:19 UT at a wavelength of 2.36 microns. The impact site has also been monitored at 2.20 microns. SPIREX will continue to monitor further impacts in these bands.

Then:

> SPIREX detected a fragment E impact site shortly after expected impact at 15:18 UT, at a comparable brightness to fragment A, at a wavelength of 2.36 microns. This data has only just become available as the South Pole communications satellites are only intermittently visible from the South Pole.

Then behind the scenes activity emerged from this message:

> The SPIREX fragment D data was significantly compromised due to the sudden onset of low blowing snow. The telescope was heroically cleared of accumulated snow by Joe Spang of the AMANDA project, and John Briggs of the ATP project, in strong winds at temperatures of -60 degrees Celsius. SPIREX will continue to monitor further impacts in these bands.

From Japan's Okayama Astrophysical Observatory. "We observed the plumes of C and D with Near-IR camera attached to 188 cm telescope."

And from Palomar Mountain: "We saw no evidence for impact B.... Impact site A became visible again at 6:20 UT, at 2.30 microns only, presumably due to a stratospheric cloud formed after the impact. No evidence of impact C was seen prior to the 200" telescope reaching its limit of 10° elevation....The night was partially cloudy, and portions of the data are badly corrupted, but there are no long gaps in coverage."

From Australia, a surprise:

> Two impacts associated with fragment C have been detected…at the Anglo-Australian Telescope at Siding Spring Observatory, Coonabarabran, Australia. The first event began at UT 06:24 17-JUL-94, approximately 38 minutes earlier than predicted, and produced a feature which is approximately 2.4" across, located 45 S, and twice as bright as the south polar cap. The second, and much stronger, event was seen at the limb at 07:20 UT. This event brightened appreciably during the first 5 minutes and then faded to the brightness of the first event after 10 minutes. Both features are still visible one hour after the initial impact. Near-infrared filter photometry of the features is now being carried out at the 2.3 m telescope, and the AAT will continue K band drift-scan spectral mapping observations at a resolving power of 300.

And so the Internet reports came through as fragment after fragment smashed into Jupiter. On July 18 I perused the reports and learned that fragment G, which struck Jupiter that morning, had been so bright that it outshone the planet in the infrared band. In its aftermath a large dark area appeared that was seen at all wavelengths at which Australian telescopes were observing. The flash lasted as long as 10 minutes. After five minutes it was so bright that it blinded the telescopes. Half an hour later that impact site appeared larger than the Great Red Spot on Jupiter. Then reports came in that fragment G had also blinded other telescopes and outshone the planet.

From the W. M. Keck Observatory in Hawaii:

The summit here in Hawaii is plagued by heavy fog. All telescopes are closed. At 21:27 HST, a minute or so before the expected impact, the [infrared telescope facility] noticed a clearing, and we opened up. At 21:39 we obtained our first frame of Jupiter, in regular K-band: a truly remarkable (saturated) plume was visible well above the limb. We started a sequence of observations at 3.4 micron: A spot was visible in our first frame, 21:40 (HST), which brightened to truly remarkable levels by 21:50, after which it decreased in intensity. At the same time the fog was coming in, and we were closing up again; we were seeing a true decrease in the spot's intensity, or whether the decrease is due to increasing cloud coverage above the telescope is not yet known.

It is raining right now; we don't expect to get any more frames tonight.

From Lick Observatory in California: "After several frustrating nights, we have finally detected the impact sites using the 36-inch Crossley telescope and our high-speed CCD system. (Is this telescope the oldest being used in this campaign? 1895.)" No doubt it was.

The telescope at the South Pole confirmed that the G fragment impact was many times brighter than all the previous events. The careful estimate of its brightness was summed up as follows: "My God, it is extremely bright." That was later judged to have been an understatement.

Japanese astronomers saw a huge dark spot larger than the red spot. Some said it was easily seen by small telescopes. And it was not long before amateur astronomers all over the world began to see the aftermath of the collisions as dark splotches on Jupiter's visible surface.

I began to think about what all this meant. Given that fragment G was supposed to have been 4.2 kilometers across, and given that it was traveling at 60 kilometers per second, its impact energy would have been about 100 million megatons of TNT, something like the K/T impactor that wiped out the dinosaurs. And there it had happened on Jupiter, in 1994! What now were the odds on it happening here? The impact produced the equivalent to 5 million Hiroshima-sized explosions going off simultaneously. Incredible! It wasn't so long ago, back in 1991 at the First International Symposium on Near-Earth Asteroids in San Juan Capistrano, California, that I had heard it predicted that we would never see objects of this size slam into any planet in our lifetimes.

A review of the bulletin board reports at 6 o'clock in the morning on Tuesday, July 19, brought in a new batch. Jupiter was severely pockmarked by impact sites and more observatories were reporting in. Sites G and H were very large, and all of the impact scars appeared to be long-lasting. A lull followed impact H.

I got up early on Wednesday, July 20, to continue my "observations" of the comet impacts through the information superhighway. The previous night I looked at Jupiter through a 36-centimeter telescope and two spots were obvious.

The first reports appear in the local paper questioning what it is that is slamming into Jupiter. Are these bits of a comet or an asteroid? Months later still no one would know.

On Thursday, July 21, Beijing checked in. They had seen the impact sites of previous events.

And so it went, round and round the world as the earth turned to give everyone a chance to see something, except those in the center of the United States, where it later turned out that two of the four impacts scheduled to occur in the appropriate time slot were duds, and the other two relatively minor.

On July 23, Reuters reported that "the comet is kaput." It was all over, and so it was, for everyone except the hundreds of astronomers who had been sharing their observations over the Internet. They would be busy analyzing their data for years to come.

A few days later, columnist Ellen Goodman remarked that "a quarter-century after setting foot on the moon, we're still faced with living with nature; not conquering it." The awesome spectacle of the SL9 impacts, for anyone who cared to dwell on the implications, was that it was there to remind us that such events do occur, that planets occasionally meet comets and asteroids in devastating collisions. It brought home to me that our concern about the likelihood of a future impact was more relevant than ever.

Columnist Howard Kleinberg wrote "Astronomers get one right." We have had our disappointments, but on the other hand those were usually blown out of proportion in advance by the media, such as we saw before the comet Kohoutek debacle, and with comet Halley. Astronomers got even those right in being able to predict arrival times.

When it was all over, I wondered what some of the observers, most of whom must have been coming down from a week-long high, felt about what they had gone through. I was given permission to send out a request over the Internet for personal responses that I might someday use in this book.

> Observers, now that it's all over, and before you pack up to leave your telescope, would you be willing to share some of your thoughts and emotions and adventures you had while observing? What is your overriding feeling as the observing draws to a close? What, if anything, did you think about earth while watching? Were there crises and dramas before or during your data taking? For example, we read that at the South Pole someone braved icy conditions to remove snow from the telescope so that observing could begin. What about the rest of you?

The response was essentially zero. Only a few hardy souls dared, and most of them had suffered set backs. From Walter Wild:

> Hello, You may or may not be interested in our efforts to observe the impact of SL9 on Jupiter at Yerkes Observatory [near Chicago]. We struggled for about two

weeks (a week prior and a week during) to get a very complex adaptive optics system…working on a 41" telescope….

Once we tackled those problems, we had clouds to contend with, so while waiting for a break in the clouds we pointed the telescope to a distant street light and closed the loop (as we say) on that to see if the system works. It did!

John Spencer from Lowell Observatory was stuck at Cerro Tololo InterAmerican Observatory in Chile and passed along this message:

No braving of blizzards at CTIO, though we did have a snowstorm one day….We are having a frustrating time here at the Cerro Tololo….I just checked the sky and the clouds are still solid and impenetrable, as they were last night, so we are having to content ourselves with working on some of the data we gathered during the first few nights of the impacts, when it was only "partially" cloudy, and observing vicariously over the Internet, listening to everyone else's reports of amazing sightings from everywhere, from the South Pole to Cambridge, England….It's almost scary to see the hammering that Jupiter is undergoing: those black clouds over the impact sites are as big as the earth, and our own planet suffers this kind of punishment from time to time. We are very fortunate to see this happening, and very fortunate that it's happening somewhere else!

From R. Stencel:

I finally came across your message on the SL9 exploder, after an intense week spent with students and postdoc help at our modest telescope atop 14,120 ft Mt. Evans, Colo. Our goal was to obtain 10 micron region spectra, and we succeeded to some extent despite monsoonal conditions. Despite the challenging working conditions (thin air, limited sleep, balky equipment), I think the students particularly will remember being part of a GLOBAL effort that will be remembered in a class with the International Geophysical Year, Sputnik, Moon Landing and Voyager at Jupiter.

From Japan:

Dear Sir, I am Junichi Watanabe, who is a chief of observations of SL9-Jupiter phenomena in [infrared] & Optical wavelength in Japan. I organized amateur-professional cooperation in Japan for monitoring SL9-Jupiter phenomena. That went through very well, and many valuable data have been taken. Moreover, we succeeded to take some valuable pictures of impacts of C, D, and K in Near-IR camera, which was developed only a few days before the SL9. I am interested in your book. Can you include such oriental excitements of SL9-Jupiter crash ? If so, please tell me what we should do. Thank you in advance.

So here it is. "Oriental excitement" was just as great as it was all over the world. The hundreds who had been successful in getting data did not think it worth sharing their thoughts and feelings. Two of the groups who took the opportunity to respond showed that there is another side to astronomy, when you go up onto the mountain and it clouds over and you pace about in frustration waiting for things to change.

All in all, Carolyn's comet produced the grand show she had hoped for. Interpretation of the data accumulated during that momentous week in July 1994 will continue for decades to come. But the event did more than provide a new type of data; it reminded us all of our vulnerability. If Jupiter could suffer so much, what about earth?

15

THE

AFTERMATH

*A*FTER all the hoopla associated with Jupiter's publicity stunt died down, planetary scientists got down to the business of analyzing their data (Figure 15-1). Simulations of the aftermath of a comet or asteroid impact had been available for years and in July 1994 many of the predictions were confirmed, albeit some more dramatically than expected. The timing of the event was almost as if to remind us to take more seriously what we have been thinking and talking about for some time.

Putting aside for a moment the implications for life on earth had something similar happened here, let's look at some of the things that were learned. Argument continues as to what actually hit Jupiter, a comet or asteroid. When the Space Telescope Science Institute sent out a press release on September 29, 1994, entitled "Hubble Observations Shed New Light on Jupiter Collision," we were led to expect an answer. The introduction gave us further hope: "Was it a comet or an asteroid?" But the institute didn't have the answer. Its observations slightly favored a cometary origin, but the asteroid possibility still could not be ruled out. Comets are mostly icy, or so we like to think, and asteroids are mostly rocky or metallic, or so we like to think. When you really get down to it, this business of the difference between comets and asteroids has launched a new cottage industry within astronomical circles.

A more recent hint that a comet was involved came from observations made

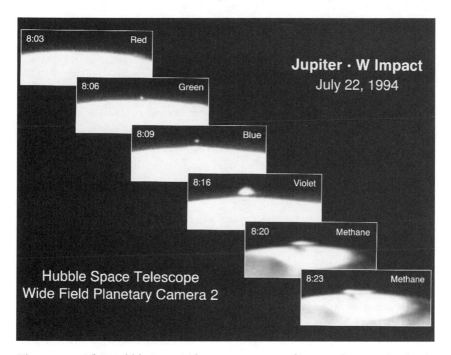

Figure 15-1 The Hubble Space Telescope sequence of images showing the development of the fireball and plume associated with the impact of fragment W taken with different filters in a time sequence lasting 20 minutes. The precursor at 8:06 is followed by the top of the fireball showing around the limb of the planet at 8:16. Then the plume develops and sinks back down (8:20) A plume this size would have encircled the earth for a similar event here. (Courtesy Hubble Space Telescope/NASA)

from on board the Kuiper Airborne Observatory, an airplane that carries a beautiful infrared telescope high above most of the water vapor in the atmosphere where it can then see more clearly. Ann Sprague and Donald Huntern from the University of Arizona and their colleagues found evidence for water minutes after two of the fragments smashed into Jupiter. The water signature, a spectral line, indicated it was at a temperature of 500 kelvins (degrees above absolute zero, or about 230 Celcius), much hotter than Jupiter's usual 200 kelvins (-73 Celcius). Although they could not rule out that the water originated deep in Jupiter's clouds, the way it came and went over a period of 20 minutes suggested that it was liberated by the impact and was part of a cometlike object.

Just before SL9 slammed into Jupiter, photographs suggested that many fragments exhibited tails, as if each was a small comet. The Hubble Space Telescope even observed individual nuclei shedding dust as they approached their end. But they did not break up further before impact. Instead, each hit with a solid wallop, which suggested a fairly substantial body was involved, more like an asteroid than a comet. By mid-1995 some consensus had developed that each fragment that struck consisted of a fairly concentrated swarm of rock, ice, and dust, cometlike structures that were, in turn, once part of a larger object that was surely more cometlike than asteroid-like.

Some of the questions essential for understanding what actually hit Jupiter will involve figuring out how deep the fragments penetrated the cloud layers before shattering, how hot the fireballs were, where the explosions occurred relative to the cloud tops, and how large the fireball and subsequent clouds of debris were. Many of them certainly covered areas much larger than earth, but what was going on inside those areas? Only the *Galileo* spacecraft was able to observe the moment of impact and its data were not transmitted for months, thanks to its failed high-power transmitter system. *Galileo*'s telescope could not see details but it did show just when the impacts occurred, which were otherwise hidden from our view on earth.

When the early reports of what happened during and after each impact were sorted out, Mike A'Hearn did what every good scientist loves to do when he or she has a first cut at exciting new data. You organize the data and then attempt to classify and label what has been observed. The SL9 impacts were no different. A *flash* was expected at the moment of entry, something akin to the meteor phenomenon when the object gets very hot and begins to burn. The flash was followed by the *fireball*, which was created by gas heated up to thousands of degrees as it is blasted out of the tunnel dug into the atmosphere by the impacting object. For the SL9 fragments, these tunnels were slanted at an angle of about 45 degrees to the horizon on Jupiter. Then a *plume* was seen which consisted of material ejected into space and falling slowly back down again. The plumes on Jupiter reached several thousand kilometers above the layer in Jupiter's atmosphere where the pressure is equivalent to that at sea level on earth. This is important, because it means that in the earth's lower gravity the material from a similar impact would rise much higher and fall back to shroud the entire planet.

Finally a *spot* or *impact site* was seen which marked the aftereffects, darkened regions in visible light, made dark by liberal amounts of dust from the shattered impactor (see Figure 15-1). On earth this "spot" would be enhanced in darkness by the dust produced from evaporating billions of tons of rock.

As with any new phenomenon discovered by scientists, most attempts to predict what will be seen before the first observations tend to be incorrect as regards details. It was widely expected that the brightest, and hence surely the largest, fragments photographed before impact would create the greatest blasts. That is not how it turned out. There was no correlation between before and after. There was, however, a mysterious inverse relationship between impact violence and how far the fragment had drifted away from the train of fragments before impact. Those fragments that were offset toward the direction of the tails had less effect on Jupiter, almost as if they were made of fluffier stuff. This did prove that the fragments differed from each other in constitution, as yet unknown.

Preliminary chemical studies indicated a host of elements and molecules, many of which must have been contained by the impacting bodies. The elements included sodium, lithium, magnesium, manganese, iron, silicon, and sulfur. Molecules included ammonia, carbon monoxide, water, hydrogen cyanide, methane, sulfur, hydrogen sulfide as well as CS, CS_2, C_2H_6, and C_2H. It was no surprise that carbon was one of the dominant elements in the molecules, because this is a property of comets in general (chapter 4). As regards molecular sulfur, S_2, this is only the second place it has ever been seen outside the earth; the other is on Jupiter's moon Io. But where did the sulfur come from? The comet fragments or from Jupiter's atmosphere?

One of the more dramatic phenomena related to the Jupiter impacts was observed by the Hubble Space Telescope. It photographed changes in the auroral patterns in the Northern Hemisphere, directly opposite the impact sites. This signified that energetic electrons had rushed along Jupiter's magnetic field to crash into the atmosphere in the Northern Hemisphere. The range of influence of Jupiter's magnetic field (its magnetosphere) is so large that if visible from the earth would be about the size of the full moon and well before any individual impact from a SL9 fragment the magnetosphere was set vibrating.

Terrestrial auroras are glowing gases created where charged particles crash into the atmosphere. Jupiter has long been known to exhibit auroras but never before had they been seen so far from its poles. Usually it is particles naturally trapped in that magnetic field that trigger the auroras as they rush back and forth in a volume of space akin to the van Allen radiation belts surrounding the earth. But when the comet fragments rushed toward Jupiter, they apparently set the magnetic field in the vicinity of the impact sites vibrating violently enough that particles were drained out of lower regions of the radiation belts to plummet into the atmosphere in the opposite hemisphere.

I found this observation on the part of the Hubble Space Telescope particularly dramatic because it carries implications for our planet. A similar auroral disturbance on earth might produce stunning power surges in transmission lines, possibly even a dreaded electromagnetic pulse (EMP). Although much of the data on EMPs, which have been produced by high-altitude nuclear explosions, is secret, it is widely known that they can blow out every electronic device not protected from the intense, varying electromagnetic field that is the hallmark of EMPs. The Jupiter impacts showed that comet fragments are capable of providing the planet's magnetosphere with a considerable kick, which suggests that the first sign that we are about to be struck by a rogue asteroid or comet might be a nationwide power outage during which all our computers are ruined.

When in 1994 I asked him about the key lessons so far learned from the SL9 impacts, Mike A'Hearn said that the data were showing, and would continue to show, how the energy of a comet impact was dispersed in Jupiter's atmosphere, which in turn tells us about what is likely to happen on earth in a similar situation. With each passing month it became more obvious that there was no solace to be found in hoping that the energy would be spent harmlessly, high in the atmosphere. A year later I sensed that this was still the key lesson revealed by SL9's violent rendezvous with Jupiter.

What was almost as wondrous as the phenomenon itself was the huge amount of time devoted to the study of the impacts by astronomers. Of course, the opportunity was not just unprecedented; it came at a time when public and scientific interest in the possibility that our planet might undergo a similar calamity had become very great,

Early in 1995 I asked Clark Chapman, who used the *Galileo* spacecraft to look at the impacts, to sum up his impressions of the SL9 collisions:

> There seems to be an increasing feeling that the impactors were relatively small. Under a kilometer in size. Not the 2 or 3 kilometers we were talking about. There seems to be a trend to believe that the dark matter was mostly cometary material rather than Jovian material. The physical models shows that very little Jovian atmosphere should have been entrained in the explosions.
>
> Another thing was that the plumes went higher than most of the models predicted. So there was more kinetic energy in the secondary ejecta. Those bright spots in the images are due to thermal radiation over areas far larger than the size of the earth but with the stratosphere heated up to a thousand degrees.
>
> You don't want to be under that. That lasts for hours and covers a territory the size of the earth. That would fry everything. The bottom line is that if something like this were to happen on earth its effects could be quite, quite substantial.

In May 1995, a special symposium of the International Astronomical Union was held in Baltimore, Maryland, to discuss what had been learned in the 10 months since impact. An excellent summary was given by Kelly Beatty and David Levy from which I drew liberally to produce the following few paragraphs. The extremely complex nature of the impact events which were observed by so many telescopes at so many different wavelengths had caused confusion until the second day of the conference when a group of space scientists lead by Clark Chapman managed to produce a cogent summary of what happened when each of the larger fragments slammed into Jupiter. It is interesting to see how much the summary has evolved from the one Mike A'Hearn offered 9 months before.

Each of the 21 fragments was not a single, well-defined object. Each was more like a small swarm of objects surrounded by a coma that was stretched out along the fragment's path over a distance of as much as several thousand kilometers. When that coma began to strike the atmosphere, a "first precursor" lasting seconds was produced which consisted of meteoric flashes that were not directly observed from *Galileo* or from the earth. Instead, they lit up some of the material still heading down toward Jupiter, and that light was seen from earth.

As more of the coma smashed into the planet, a vivid meteor shower was generated. Then the incoming fragment disappeared behind Jupiter's limb and at this point, *Galileo's* cameras recorded the flash of first entry. Some earth-based telescopes detected a concurrent, faint flash signifying what had happened out of sight around the limb. Light from this entry flash appears to have bounced off trailing dust in the coma which was also seen from earth.

The next phase was the "main fireball," an enormous bubble of superheated gas that blew back out through Jupiter's stratosphere, along the tunnel the fragment created upon entry. The fireball heated gas to over 10,000 kelvins, twice as hot as the surface of the sun. This phase was monitored by *Galileo*.

A minute later telescopes on earth saw a bright flare as the fireball emerged from behind Jupiter's limb. This the Chapman group called the "second precursor." Based on timing of the event, the fireball blew outward at 17 kilometers per second (or about 38,000 mile per hour). As it ballooned into space it cooled and faded from *Galileo's* sensors, while infrared telescopes on earth were able to watch it slowly fade.

The final phase, which they called the "main event," was produced up to six minutes later by the plume of material that fell back at 5 kilometers per second to heat vast tracts of the stratosphere to 2,000 kelvins. "This unexpected 'splashback' generated a burst of infrared energy at least 1,000 times brighter than the first precursor, and it saturated the sensitive cameras on many terrestrial telescopes." It appears that some telescopes detected secondary flashes, which may have been produced by the plume bouncing and then falling back down again.

Then the flattened remnants of the plume (see Figure 15-2) spread outward to produce the dark stains that were so widely reported. These were dark in visible light, due to absorption of sunlight by the solid matter of which they were made, but bright in infrared because of heat radiation from the sun reflecting off the dust within the markings.

Over the next weeks and months, the dark dust was caught up by high-speed winds at Jupiter's cloud tops and began to spread out to become tracers of wind patterns (Figure 15-2). What were once dark rings of material stretched out to become "curly-cue" features. Although individual impact sites were still visible a month later, no long-lasting effects in Jupiter's structure had been expected. Nine months after the impacts, Jupiter still bore the impact scars as a band of enhanced infrared emission where the material from the 21 fragments had dispersed all the way around the planet.

Many more details about the impacts and the way shock waves traveled away from the impact sites to trigger molecule formation, for example, have been learned, but discussion of these details is beyond our scope. Most of the data are

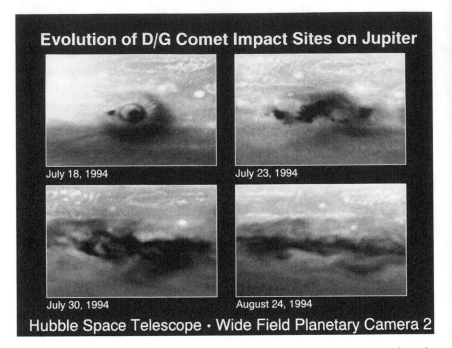

Figure 15-2 The evolution of the D/G impacts sites on Jupiter following the collisions of July 17 and 18, 1994. Comet fragment D created the small dark spot at the left in the first frame. Fragment G was a very violent impact that created the circular ring structure, probably the "sonic boom" from the blast. The outer-ring feature shows where dust from the disintegrated comet fragment fell back onto Jupiter's atmosphere. In the first frame it covers an area larger than the earth. Over the next month the dark debris is swept into a long streamer by Jovian winds. (Courtesy H. Hammel, MIT and NASA)

quite technical and will enable the experts to obtain an even more accurate picture of what happened. For example, the studies of the aftermath may yet reveal a great deal about the delivery of key elements and molecules to the earth by comets, not only in recent times but when the planet was still taking shape. This information will be very interesting to those who study the origin and evolution of life. The SL9 impacts were about as close to a series of laboratory experiments that anyone could have dreamed up to discover what happens when a comet slams into a planet to deliver its load of water, dust, rocks, and molecular ingredients from the depths of space.

In retrospect, it is apparent that the most important lesson learned from the Jupiter impacts is that they contained little good news for those who might be inclined to downplay the potential for catastrophe resulting from similar impacts on earth. To the outside observer who has listened to the prognostications about what size object is likely to wipe out civilization, the Jupiter news is nothing short of alarming. A few years ago it could still be argued that the comet or asteroid would have to be at least 1 kilometer across to do serious harm. Impacts by objects of this size were believed to occur once every few hundred thousand years. The SL9 fragments that struck Jupiter were more like a half kilometer in size and they produced fireballs followed by plumes of hot matter that would have girdled the earth. This suggests that the potential for planetwide chaos, should one strike here, may have been underestimated. These smaller objects are likely to smash into the earth every 20,000 years, on average.

The overriding message of the Jupiter impacts is that planets continue to be struck by very large objects from space, so large that they can produce devastation that can be seen for months afterward, even across interplanetary distances. In July 1994 we could peer through our telescopes at Jupiter in all its remoteness (800 million kilometers away) and marvel at the amazing rings of dark material debris from the comet/asteroid fragments, debris that blackened the skies of Jupiter over an area several times as large as the earth. And that was done by each of half a dozen fragments. For the past fifteen years, computer simulations of such an impact predicted that the world would become enshrouded in a thick blanket of debris, soot, smoke, and dust were it to happen here. The Jupiter impacts confirmed this. The earth would have been lost inside the dark markings that were the signatures of the collisions on Jupiter.

Strangely, none of this was hinted at by the participants at the press briefings during the week of the impacts. Was it that they were so excited by the abstract nature of their observations and their scientific data that they forgot that these impacts signified, above all, that we, too, are vulnerable? Or was it that they dared not step out of their role of sober, cautious scientists to speak of what might happen here some day? There was no mention that if something like fragment G had struck the earth, people would have perished by the billions and that the fabric of civilization would have been destroyed. No one mentioned that following such an impact a great blanket of dust would spread through the stratosphere and that as a result sunlight would be totally cut off for months. Not a hint was offered that photosynthesis would cease and global crop failures would result. Nor was mention made of a global freeze that would follow the collision, the so-called impact winter. It was almost as if the participants in these news conferences had agreed in advance not to paint too horrid a picture of the consequences of a terrestrial impact.

The SL9-Jupiter collision came at an extraordinarily opportune time. It was seen only 14 years after the possibility for catastrophic collisions between comets and earth first entered the mainstream of scientific thinking. It was timed so that a spacecraft, *Galileo*, just happened to be heading for Jupiter, thus giving space scientists a direct view of an impact that occurred behind Jupiter's limb as seen from earth. And it happened that we had a major telescope in orbit to obtain a clear view, the Hubble Space Telescope which had just been repaired. Finally, the impacts that occurred on Jupiter offered the sort of experiments required to test theoretical calculations that had been made over the past decade as to what happens when a comet smashes into a planet.

SL9's rendezvous with Jupiter could hardly have come at a better time. One cannot help thinking that we were really very, very lucky to see this now.

16

THE SEARCH

*T*HE earth orbits the sun in a veritable swarm of asteroids that have the
nasty habit of occasionally slamming
into the planet (Figure 16-1). To add to the potential danger, comets sometimes
wander into the vicinity of the sun, break up, change course, and hang about
posing a threat for up to tens of thousands of years. We don't really want to be
hit by any of these, so what can we do to avoid the blows? Most important, we
have to keep our collective, astronomical eyes open and try to spot the dangerous ones before they get here.

With almost religious fervor some planetary scientists have been seeking
near-earth asteroids in recent years to determine how many may be out there
that might yet pose a threat to our planet. These scientists have met on many different occasions in the past five years to discuss search strategies and what to do
next. The first major meeting of this type, open to more than those in the inner
circle that grew from the discovery of the iridium layer in the K/T clays, was
held at San Juan Capistrano in 1991. It was billed as the First International Symposium on Near-Earth Asteroids and brought together interested scientists from
all over the world.

Eugene Shoemaker was there and he urged caution about describing what a
civilization-destroying asteroid is. After all, two out of three will hit water, he
said, ignoring the point raised in chapter 12, that a water impact is almost certain to produce a greater catastrophe than a land strike, especially if several fragments should be involved. He estimated a civilization-threatening impact once
every few million years. Tom Gehrels of Spacewatch at the University of Arizona

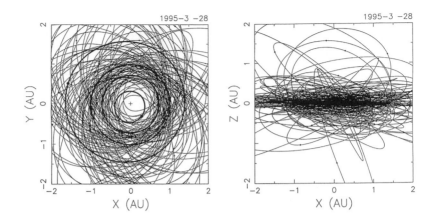

Figure 16-1 The orbits of 115 near-earth asteroids plotted for March 28, 1995, by Syuzo Isobe, National Astronomical Observatory, Tokyo, Japan. The plan view on the left shows orbits superimposed on those of Mercury, Venus, earth, and Mars in heavy circles. This makes it dramatically obvious that the earth orbits the sun through an asteroid belt. The diagram on the right is a side view of the NEA orbits with the location of some of the asteroids more clearly indicated for that date. The earth's orbit extends from -1 to 1 AU. (Courtesy Syuzo Isobe and Makoto Yoshikawa)

did not take well to this caution and pointed out that "Humanity is ill-advised to say the probabilities are so low as to ignore them. I will not take that point of view."

As I listened to the talks and arguments, a sobering thought crossed my mind. Resting upon the efforts of this group of searchers might be the future of humankind. All of them realized that estimating probabilities of that magnitude was nit-picking. The real issue was that our place in space is not as safe as we would like it to be. In fact, we live in a very unsafe environment, no matter how benign it may seem most days.

Eugene Shoemaker said he was involved in the search not to avoid collision but because he thinks that asteroids present the next and only logical step into space. "These objects represent stepping stones into deeper space," he says. "Instead of going around and around the world getting nowhere, let's go to an asteroid. That's how we'll get to the moon. Not by dinking about in space stations."

He stressed that if one wants to go exploring and doing science in space then one should do so and not pretend that the space station is any use. "We've got to land somewhere and do science there. In an exploratory mode we can learn about the science as we go." He argued that a hands-on approach will teach humankind about the real difficulties of space exploration. "Landing on an asteroid is a great field problem in how to study rocks."

He made a lot of sense; there is little or no point in hanging about in space stations. That's something like Columbus deciding that instead of sailing west he would spend a few years sailing around the bay to make sure that his sailors could deal with seasickness and that food supplies had been accurately estimated. Perhaps the owners of the ship would rest more easy if they could see their investment out in the bay. If anything went wrong, they could at least rescue the sailors. But once they let Columbus out of sight, who knew whether he would come back? The point is that no one did know, and perhaps that is what is at issue here. As regards the exploration of space, I have for some time thought that the United States has lost its will to take risks, for that is what exploration requires.

Most of the great sagas of exploration in the Western world sponsored by governments or royalty originated in Europe and Great Britain, the opening up of the West notwithstanding. The latter was a venture of individuals willing to take the risks for themselves.

Without risking his life and that of his crew, Columbus could not have set sail. We might do well to recognize that the true heroes of exploration are those who risk all. But the psychology of the U.S. Space Program has become to risk nothing. Caution may have been carried to an extreme, a caution exemplified by keeping astronauts in safe orbit (and close to earth in a space station?) and pretending that that is the way to space. "Dinking about in a space station" avoids the issue. It creates an illusion that we are doing something, which can be justified on endless grounds, but this is an end in itself and gets us nowhere.

The issue of landing on and exploring asteroids is a key to whether we will ever be able to deflect an object that is headed toward earth (chapter 17). It is well nigh impossible to plan to deflect an intruder if we do not have any idea what it is made of; whether it is a rocky object, a dormant comet, or some fluffy snowball of the sort that broke up near Jupiter and crashed to its spectacular demise in July 1994.

No one at the San Juan Capistrano meeting could even begin to consider the issue of what we could do if we found an asteroid that was predicted to collide in a year or so. No one was ready to take that step. As scientists, the first step is to obtain more data so that we can be as well informed as possible. Then we predict what happens next, and based on those predictions plan a rational policy of continued monitoring and possible preparation for a rendezvous. For example, if we had to intercept an asteroid on an emergency basis, it is essential that we know what an asteroid really looks like, up close. If it is a dead comet consisting of an outer crust protecting masses of ice and gas, for example, then the techniques for deflecting it might be quite different from those used for solid objects. We do not want to be placed in an emergency mode and find that our efforts to deflect an asteroid result in its disintegration so that, instead of one large object, hundreds of small ones pepper our globe.

During the meeting we gathered for a special session to hear about the reports being presented to NASA concerning the detection of NEAs. The group huddled around an overhead projector, seats were pulled out of alignment, and the audience was rapt. Serious scientists were discussing what could be done to avoid the extinction of the human race. None of them argued about whether it might or might not happen. That was no longer an issue. It will happen, some day. It could be tomorrow, or 100,000 years from now. But even this uncertainty pales before the great dilemma that confronts us. Do we wish to gamble with the continued existence of *Homo sapiens*? NASA had asked for their expert opinion, and these experts had offered their first thoughts on the subject. Their conclusion: "We must act."

A Russian scientist took the opportunity to quote a remark once made by President Reagan: "Your President said that what we needed to unite the nations of the world was a common enemy from space." When he said it, Reagan was, of course, hoping that a Hollywood-type extraterrestrial would threaten us, but the point was well taken. "Now we have a common enemy," the Russian went on. "Let us work together to fight the asteroids. This is a great opportunity to work together." The San Juan Capistrano meeting ended with little accomplished beyond the sharing of ideas.

In 1995 I visited Tom Gehrels who runs the Spacewatch project at the University of Arizona which uses a 36-inch (81-centimeter) telescope to search for asteroids from atop Kitt Peak outside Tucson. He gave me an update of the progress they had made. "We are finding about 35 especially interesting objects per year. This is steadily increasing." Their total reached 100 in 1995.

I asked him where he now thought the real danger lay. Was it with random objects approaching earth, or from objects in the Taurid Stream? "I think that the Taurid hazards are real," he replied. "It is difficult to convince people that it is real. It is a psychological issue; why is it so difficult to convince colleagues? It's a silly situation right now. All these astronomers talk about all the work they are doing on all their telescopes [studying distant reaches of space beyond the solar system], and all the theoretical work they are doing, but one of these days they may wake up and find that there is no *Astrophysical Journal* anymore to publish their great discoveries and that they themselves may be dead."

He added "The [coherent catastrophism] proposition has to be taken seriously. I am very grateful that [the Musketeers] have pushed it so hard. And a lot of the credit goes to Clube and Napier who have been talking about this. I think it is very good that they have done that. I think it is a healthy debate that is going on. Its a matter of the numbers."

Just what are those numbers? "The uncertainties you are referring to are systematically expressed as follows. With the present efforts with Spacewatch, it will take 50 years or more before we know the ones that can destroy our society. The [asteroid size] boundary is somewhere around 1 kilometer, depending on its composition. With a metallic object it can be as small as two tenths of a kilometer. That will cause a global disaster." Asteroid 1989 FC was in that size category.

He admits that even he had never fully appreciated the potential danger, even though he had shown a table in his book, *On The Glassy Sea*, summarizing collision probabilities years before. "I have been showing that table since 1975 and I really didn't believe it, and became convinced of the *horror* of the situation of the *stupidity* of the situation, and the astronomical responsibility of the situation until a few years ago. Why was I so slow?" He admitted that he had been living with this and had not appreciated the magnitude of the risk. In recent years he has changed his mind and is now one of the most active in discussing the threat and doing something about it. "The first step is to find the dangerous objects," he said, which is what his Spacewatch program is doing.

"How could I expect my astronomical colleagues to be any faster than that?" he asked rhetorically. "But what I really would like is that all telescopes that would be suitable would be made available for this, instead of this puttering around we've been doing. Spacewatch and Palomar and a little bit here and there, which will take 50 years before we will have removed the uncertainty." He stressed that a concerted, worldwide effort would reveal the full census of the current NEAs, all 1800 of them.

"The chance is 1 in 10,000 that there is now one on its way to the Earth." That uncertainty can be reduced to 1 in 100,000 by concerted effort, which will, in some sense, allow us to feel 10 times safer, or at least get on with the job of avoiding an impending impact.

"There are 20 telescopes that can be used right away," he added. However, those telescopes were being used by astronomers for other research programs, and none of them would be willing to sacrifice their valuable observing time devoted to studying distant galaxies or stars for the benefit of near-earth asteroid searches, not unless some absolute dictator decreed that it should be so.

I asked about the Air Force telescopes that might be made available for the searches. "No, no, no. That is part of this piddling effort that is going on right now. I'm talking about all suitable telescopes at all observatories under all existing budgets of people salaried to just get on with their astronomy." They would devote a few years of dedicated effort and perhaps their telescopes would be called upon to obtain some follow-up observations. "This would be an effort all around the world at existing observatories using only *some* of their telescopes."

"If that were done the uncertainty would be removed within the next five years, which reduces the chance of us being eliminated [by asteroid impact] from one in 10,000 to one in 100,000. Now you can argue, and astronomers will do that I am sure, that it is still a very relative matter."

Spacewatch, like the other search efforts around the world, is terribly strapped for money. Even though the danger of impact was becoming more widely appreciated, that has not helped it obtain funding. "More and more people are calling in, and it is penetrating to the schools. And its all getting into the textbooks. The new textbooks now have the proper story in there, of how the dinosaurs got eliminated. This takes time. It grows and will become familiar to all and that's happening quite rapidly. Somewhere along the line you may get taxpayers to put a little money into this. But I don't see that happening at this time at all. I think that the Shoemaker Committee will not help Spacewatch at all." (A brief summary of this committee's report, commissioned through NASA by the U.S. Congress, is given at the end of this chapter.)

"Out of all the publicity [in the past] I only got a budget cut. Now that we are doing so well I got an 18 percent budget cut. It is a serious situation."

During my visit I got to see what the search for asteroids is really like. I drove out of town, toward Kitt Peak which loomed in the distance, shadows highlighting valleys cut through millions of years of erosion, blue skies, cirrus clouds behind the mountain. From afar, you can see an occasional dome silhouetted at the mountain top. It is out here, above the desert splattered with saguaro cactus, that a few astronomers scan the skies in order to help protect the earth.

The winding trek up the mountain takes you to over 2000 meters, and as I drove a local radio station played Tchaikowsky's *Tempest*, a fitting fanfare. At the visitor area atop the mountain, a 14-inch (35-centimeter) telescope was set up to point at the first-quarter moon. I looked through it and saw impact craters dramatically visible, a stark reminder of our past and our future.

Robert Jedicke met me. He was a relative newcomer to the team, a former particle physicist drawn by the lure of the mountain, and would be observing that night, from sunset to sunrise. We began to talk over dinner, known as breakfast to the observers on the mountain. When they leave for their telescopes, they pick up their night lunch on the way out of the cafeteria.

How many NEAs has he discovered? "If you include the one last night, which is beginning to look like an Amor, I think it is 15." It was indeed a Mars crosser. They can tell from the extent of the trail on the computer screen that displays the images collected by the CCD (charge-coupled device, an electronic sensor) attached to the 36-inch telescope they use. The asteroids look starlike but move between images. The really nearby ones are actually streaked (see Figure 9-1). "We take three images each lasting 30 minutes of the same area of sky and the software almost in real-time looks for starlike objects that have changed their position."

I asked him what his personal feeling was about how seriously we must take the NEA threat. "That's a difficult question," he replies. "You cannot deny the odds, the statistics. It is a very important topic. In the next hundred years, what are the odds that we will have a global catastrophe caused by an impact? And what are the odds that we are going to have a global catastrophe caused by disease, pestilence, environmental hazards, or something man-made. The odds of those are larger than those of an impact. I don't deny the fact that NEAs are a hazard, but if you look over the next 100 years there are other hazards that are more important."

We entered the telescope control room, a constrained area not for claustrophobics, packed with equipment, books, computers, and an old armchair too cluttered for anyone to sit on. First he had to top up the CCD on the telescope with liquid nitrogen. On the way we took a moment to pause atop the dome to look out over the desert and watch as the shadow of the mountain moved toward the distant horizon as the sun sank at the end of a gloriously clear day.

The process of filling the detector involved using a low-tech, cost-effective broomstick to support the container from which the liquid nitrogen was poured into the apparatus. The broomstick was their proud symbol of independence, of working with what they could get.

It turned out that two nights before I visited, Jedicke had apparently found a new comet, but he was not yet certain it was real. He had experienced many false alarms, he said, but this one looked good. Would it be seen that night, suitably shifted on the sky? "I don't know. Hopefully we will know soon."

Back in the control room he asked me to avert my eyes so that he could surprise me with something he'd display on the screen. When I looked I saw an asteroid jumping back and forth between two images blinked by the computer. It was a spectacular sight, a little point of light jumping between the fixed stars. The experience sent chills up and down my spine. Was that how our species would receive its first warning that the end of civilization might be at hand? "It's moving about a quarter a degree a day," he explained, bringing me back down to earth.

Then he called up data from the previous night to show me the comet he found (Figure 16-2). When I first saw this," he said, "it was superposed on that star, which gave it a tail. I see lots of things like that which I think are comets. That's why when I first saw it I wasn't too excited." He explained that the tail was too faint to be associated with a comet if it were as bright as the star that was located there. "I thought maybe it's just some sort of a weird meteor. But I noted it. And I looked out for it when it reappeared [in the next data set]. In fact I was writing a letter to my brother on e-mail at the time, and I typed '[expletive deleted], I've just discovered a comet!'"

"When our automated software finds a comet, we've agreed that the comet will be called comet Spacewatch. But because it didn't pick it up I am the sole discoverer."

"Does that mean it will be called comet Jedicke?" I asked.

"Yes, it will be called comet Jedicke." Yet he was not sure because there has been talk of changing how comets were named.

As the observing proceeded, we watched the images of the star field build up on the screen. Suddenly a ringlike object appeared. "A planetary nebula, proba-

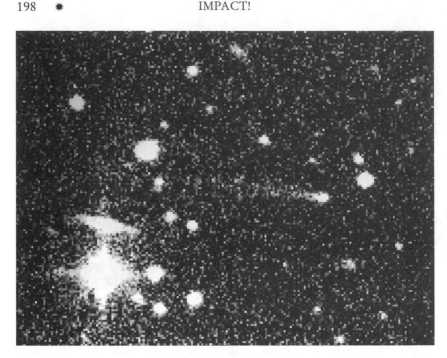

Figure 16-2 Comet Jedicke also known as 1995A1 as seen in the discovery data obtained as the comet was departing the solar system beyond Jupiter. This faint object, the streak in the center of the image, never came close to the sun and was found during the Spacewatch search for near-earth asteroids using the University of Arizona's 36-inch telescope on Kitt Peak. (Courtesy Spacewatch, Lunar and Planetary Laboratory, University of Arizona)

bly not a well-known one. This one is so faint it won't be in any regular planetary nebula catalog." I notice how beautifully symmetrical it was.

The software draws attention to objects that might be streaks and it displayed their images. It was up to the observers to decide which ones to save for later study. Most of the streaked images were faint galaxies, however, of little interest to them.

A vivid streak of light suddenly illuminated the screen "This is a meteor," he said.

"Wow!" I responded profoundly to illustrate how stunned I was.

While we waited for the comet field to come into view, Jedicke elaborated on the program. "Spacewatch is finding about 2500 main-belt asteroids per month. Eighty percent have not been cataloged before. For each month where we find 2500, we find 2 or 3 NEAs. Our yearly total is about 30 to 35. We're working toward increasing our efficiency." He suspected that they might be loosing 40 percent of the objects which they could, in principle, detect.

"We're trying to find asteroids in real time in a noisy background. That's difficult. The program will mess up sometimes."

He was not optimistic about the future of the search programs, worldwide. "I am very worried. I am still very young in this field. I have a bad feeling about all these projects. The general trend is to cut back on funding. So, is there a future in this business? I really hope this all works out."

He looked at the screen and gestured with a pencil. "On a clear night, under good seeing, we might find 100 to 200 asteroids on a single scan like this." Since they did about 6 scans a night, data handling was a major part of their study.

Finally the telescope was pointed toward where the comet should be. Within minutes we both saw it. A few days later, its discovery was announced to the world and it was called comet Jedicke. It never came very close to the sun; when he found it, it was already headed out of the solar system.

It was concerns about the threat of NEAs that in part motivated the likes of Gene and Carolyn Shoemaker, Tom Gehrels, Eleanor Helin, as well as Duncan Steel in Australia with his colleagues, to step up searches for NEAs. They appreciate that no one can begin to make coherent long-term plans for the future of civilization without knowing what's out there posing a threat. What these searches have found has come as something of a shock. There may be far more large asteroids in the *Apollo* family of earth crossers than anyone dreamed a decade ago.

Asteroid searching produces a better census of NEAs and also adds to the list of main-belt asteroids between the orbits of Mars and Jupiter, well out of harm's way. Small objects are apparently evaporating from the main belt—that is, they are in unstable, chaotic orbits. Others leave the main belt following collisions among themselves. The main belt is therefore spawning earth crossers.

In 1993, 6000 asteroid search observations netted 600 main-belt objects per month. That's already a lot. Most of those had been discovered previously, but that still implied a lot of data analysis. If one or more of the planned NEA surveys goes ahead, a bonus of 100 to 1000 times as many main-belt asteroid detections will be obtained. It is estimated that the number of 0.1 kilometer objects waiting to be detected in the main-belt asteroid population may range from 10 billion to a trillion.

Three surveys are currently underway. They are Spacewatch run by Tom Gehrels in Arizona, the Planet Crossing Asteroid Survey (PCAS) at Palomar Mountain in California involving the Shoemakers and Eleanor Helin, and the Anglo-Australian Near Earth Asteroid Survey (AANEAS) led by Duncan Steel.

Six more are being planned, including a more comprehensive one in Australia, the Global Electro-Optical Space Surveillance (GEOSS) using Air Force telescopes started under the guidance of Eleanor Helin, LONEAS at Lowell Observatory, a survey using a telescope on the Côte d'Azur in France, and Spacewatch with a larger telescope. At this time these programs are all hopelessly underfunded and being threatened with budget cutbacks. (Unlike professional athletes, NEA searchers cannot go on strike to demand more money!)

Edward Bowell of Lowell Observatory has described what might be expected of the new searches in the very near future. Soon the whole dark sky will be observed every month and the planned surveys will produce between 30,000 and 260,000 asteroid detections per month! Although the searches are optimized for NEAs, they cannot help but pick up thousands upon thousands of main-belt asteroids. If so many objects are going to be watched scooting across the sky (Figures 9-1 and 16-2), how will all these data be analyzed and interpreted?

No one really knows what headaches will be created by large-scale asteroid search programs. One thing is certain; the old ways of doing business will have to change. Until now, if you found a new NEA or a main-belt asteroid you could name it. At the very least, you announced the discovery with your name attached and that counted toward the list of publications so important for scientists in the quest for promotion. But soon those involved in the searches, and those who pay their salaries, will need to rethink what gets published and how anyone gets credit.

In the upcoming bonanza of discoveries, it is expected that false detections will be made at 500 to 1000 per hour. Repeat observations will sort these out but even that means that 500 objects per hour will have to be observed for an hour, which also implies that initial orbital calculations will have to be produced at a rate of 500 per hour. All this work will lead to actual orbit determinations having to be made at the rate of 20 to 200 per hour, which implies an almost unbelievable amount of work for those involved in the searches, even if all the drudge is taken out by using large computers to look initially at the data. Duncan Steel foresees that the only way to handle the work will be to organize a military style operation in which data handling becomes highly regimented.

The point of asteroid search programs is to discover if among the thousands of objects spotted every month there might be potential earth impactors. Monitoring such objects must then continue in order to detect changes in their orbits because an object that looks safe this year may yet become a threat in 50 years time as the result of gravitational encounters with planets or the sun.

In early 1995, Jack Hills and Peter Leonard of the Los Alamos National Laboratory in New Mexico published a study of what was likely to be seen in the last weeks before an asteroid strikes earth. Rather surprisingly, the asteroid would not appear to move very much across the sky and might be very difficult to detect, or, if it were detected, it might at first be thought to be an object in a more distant orbit. The problem is that an object about to strike earth would tend to be moving more or less toward us and would exhibit relatively little sideways motion across the sky. Sideways-moving objects are the ones Spacewatch and other NEA searchers are looking for.

The danger from a potential impactor might only be spotted by looking at the object from several different sites on earth simultaneously; then its large parallax would be observed. (Parallax is the apparent change in the position of an object with respect to a distant background when viewed from slightly different locations.) That would give the asteroid's distance and from hour to hour that parallax would change, signaling that the object was coming nearer. Its motion across the sky, however, would be small and, for some directions of approach, might even be close to zero.

Hills and Leonard wrote that "Impacting asteroids greater than 100-meters in diameter, which is near the minimum size that produces significant local damage, appear as stars of at least visual magnitude 18 during their final 10 days to impact unless they approach earth from the vicinity of the sun in the sky."

They note that such objects will be readily detectable with even small telescopes in the range 5- to 16-inch (12 to 40 centimeters) equipped with CCDs. They propose that a station for detecting such asteroids would have 20 telescopes of this size and one with an aperture of 36 inches (90 centimeters). Data would have to be immediately compared in real time to determine if any object is on a collision path. This project might even be done by amateurs with suitable telescopes. The pleasure from becoming involved in this type of search would be to watch some close passes. The downside will come when the object does not miss.

As Hills and Leonard point out, the air blast from stony objects as small as 60 meters will destroy cities. When this description is given, we must not forget that such a blast would kill millions. If a stony object 200 meters in size crashes into the ground, it would do severe regional destruction, and according to most guesses may do planetwide damage. If it should hit an ocean, the damage from tsunamis could be even greater. "A warning time of a week would be sufficient for a single rocket equipped with a nuclear explosive (using existing rocket boosters and nuclear explosives) to deflect from earth impact of an asteroid with a diameter of up to 1–2 kilometers if such a launch vehicle were on standby alert for such a deflection." The warning time could even be used to evacuate an impact area or coastline where tsunamis might be expected.

Hills and Leonard make a fascinating suggestion. They estimate that what is needed to provide continuous coverage to find potential impactors is six sites around the world each equipped with 20 small telescopes (from 12 to 40 centimeters in diameter) to cover all the sky farther than 40 degrees from the sun. A few sites would have a 90-centimeter telescope to peer closer to our star. They estimate that if this project had to be funded from scratch it would cost about $3.5 million. Since most of this cost would be in the CCD detectors and computer workstations, each of which is falling rapidly in price, they think that the cost of equipping a site with the necessary hardware, other than the telescopes, will soon approach that of an automatic weather station. As they conclude, "the long-term cost effective approach may be to put an automatic detection site at all the major observatories of the world" where they could be maintained by the local staff.

Problems arise during moonlit nights when the earth could still be unexpectedly hit. Also, when looking very close to the sun, infrared telescopes would be needed to improve the chances for detection. If we are to take this seriously, a telescope orbiting Venus would be even more useful for keeping a close lookout for earth.

In the summer of 1995 the challenge of finding the earth-threatening asteroids received a tremendous boost by the publication of a high-level report given to NASA by the Near-Earth Objects Survey Working group under the chairmanship of Eugene Shoemaker of Lowell Observatory in Arizona. It had been given the task to "develop a program plan to discover, characterize and catalog, within 10 years (to the extent practicable), the potentially threatening comets and asteroids larger than 1 kilometer in diameter." NASA, in turn, was acting on direction given by the Committee on Science, Space, and Technology of the U.S. House of Representatives, which had been galvanized into action following the drama of the comet impacts on Jupiter in July 1994 (chapter 14).

The Shoemaker Committee report presents plans to achieve these goals, specifically for finding 60 to 70 percent of the objects larger than 1 kilometer diameter. These are potentially civilization-destroying, and there are thought to be about 2,000 in earth-crossing orbits. This crucial project would only cost about $5 million per year for the first four years and then decrease to $3.5 million per year.

If the U.S. Air Force were to cooperate in the search, a 90 percent success rate might be obtained by 2006. The point is that "The Department of Defense has the responsibility for protecting the United States against catastrophic losses," and there would no loss greater than would follow an impact of a 1-kilometer object anywhere on earth. Apparently the Air Force Space Command is in the process of evaluating whether its technologies should be improved and shared for the defense of the planet. (The committee was unable to estimate the costs for the Air Force contribution.)

The report further urged international cooperation in this crucial venture and pointed out that the use of existing telescopes would allow this project to get into full swing very quickly.

Although the task of finding most of the 1-kilometer-sized objects in 10 years was very practical, the further goal of making the census 100 percent complete "would require a Herculean" effort. Furthermore, finding the 100-meter objects capable of producing vast damage on a regional scale (see chapter 13), in particular in the shape of tsunamis around the shorelines of an ocean, will be a far greater project still, one that, I suspect, might take a century or more of constant vigilance.

The Shoemaker Committee specifically dealt with the threat of a global catastrophe; the potential for regional catastrophe will be something we will have to live with for a very long time to come (one hopes!). But every project must start somewhere, and the Shoemaker Committee report offers a realistic, cost-effective manner in which to proceed. I, for one, hope that there will be no delay in carrying forward the committee's recommendations

Once an earth-crossing asteroid is seen to be on a collision course, then what do we do? That depends on how much advance notice we are fortunate to have.

17

DODGING THE

ASTEROIDS?

*F*INDING asteroids and comets that may someday slam into our planet is the first step. What do we do then? This question is being given a whole lot of attention. In early 1993 NASA and the U.S. Congress received a report of the Near-Earth-Objects Interception Workshop (Spaceguard), the first step toward creating a program for pushing aside approaching asteroids. The report stated that "There is a clear need for continuing national and international scientific investigation and political leadership to establish a successful and broadly acceptable policy."

There are two or three options open to us to avoid being wiped out. The first is to step out of the way. This may not sound very practical, and it isn't, at least not for a planet-load of people. However, if we plan ahead we could ship a few thousand human beings to other parts of the solar system so that if the earth were to be struck, they, at least, would survive. This would only be a privilege for a few, and getting back to earth after the cataclysm could be a rather large problem in itself. Who will welcome them back upon their return? Where would they land? If we could afford to set up colonies on the moon or Mars, the colonists could wait until after the dust had settled before attempting to return. The problem with this option is that, after a really healthy thwack, the earth's environment would be so altered that returning human beings might find this to be an alien planet.

The second way in which we could avoid getting hit would be to place an object between the onrushing comet or asteroid and ourselves. For such an emergency it might pay to place a few asteroids in geocentric orbit to be maneuvered when we need them. Then we could watch the spectacle as one asteroid slams into another, possibly showering the planet with small bits of debris that might do no more than create a spectacular display of fireballs—if we get it right, of course.

The third option is being seriously discussed in many quarters, especially where the threat of job cuts is affecting those who used to spend time planning to defend the United States as part of the so-called Star Wars program, the Strategic Defense Initiative or SDI. This involves either pushing the threatening object aside, or shattering it out in space.

These options will only work if we see a rogue asteroid or comet well before it gets here. That is why the business of NEA (or NEO, a more general term that includes all near-earth objects, asteroids, and comets) searches is regarded by so many as such an important topic. We need to learn a great deal more about what's out there before we can intelligently plan to avoid impacts. As Lucy McFadden at the University of Maryland recently said, "Every time we study near-earth asteroids, we're constantly surprised. The more we look, the more we find and the more we see new types of asteroid populations."

Consider the statistics of the NEAs. As of mid-1995 there were about 350 known, more than half of them discovered in the previous five years. It is estimated that there may be as many as 2,000 in the 1-kilometer-size range, fatal for civilization if they were to collide with the earth, each one capable of blasting us back into another dark age.

If you prefer to regard a half kilometer object as potentially lethal, and something this large is certainly capable of wiping out a few billion people especially if one splashed into the Pacific Ocean, there may be as many as 5,000 to 10,000 of them in earth-crossing orbits with an impact expected on average every 5,000 years or so (chapter 13).

The conclusion of the Spaceguard report was that for the smaller objects, those that could produce incalculable damage to civilization, deflection is an option. This will almost certainly require the detonation of nuclear devices close to or even on, the approaching asteroid. Such a program can only be carried out after a great deal of study and would have to proceed in the context of concern about having nuclear devices blasting off from the surface of the earth. It would be ironic if in the final analysis ill-informed opponents of a plan to deflect a rogue asteroid were to obtain a restraining order on the launch, in which case one of the last headlines ever to be published in the world might read "Court Prevents Launch of Asteroid Bomb."

On the serious side, plans for asteroid interception will have to proceed in an atmosphere of international trust, because it may turn out that the design of the nuclear device that would be required to deflect the object in its course would have to be tested, preferably on an otherwise harmless asteroid. Such tests might also be useful for redirecting a few into storage orbits, later to be used as shields, although there are dangers with such actions.

There are several scenarios as regards our ability to avoid a catastrophe. The most optimistic is that we would have decades of advance warning. In this case it would be possible to impart energy to the asteroid or comet at perihelion, its point of closest approach to the sun. This is the most favorable time to add a small amount of energy to make the biggest difference to its orbit. It would be next to useless, and far too late, to try to impart energy while it was heading straight at us. The best thing that could then be done would be to try to blow it up and suffer the rain of fragments that would then strike the atmosphere.

For comets, a year's advance warning might be all we'd have and then a high-energy launch would have to be undertaken as a desperate last measure. A high-energy launch refers to the spacecraft having to intercept the object at a very high velocity.

The worst-case scenario would involve only a few weeks' or days' warning, in which case nothing could be done except duck. At such times the entrances to caves might become prime real estate, although there would be little to guarantee one's safety once inside, given that violent earthquakes following a severe impact might cause them to collapse.

As regards the imparting of energy, standoff nuclear explosions are thought to be the most reliable and efficient. The theory of a standoff explosion is elegant. An intercepting spacecraft carrying the nuclear device moves parallel to the asteroid and the explosion creates so much energy, in particular in the form of fast-moving neutrons, that the surface of the asteroid on the side of the explosion is heated slightly and gas and dust begin to stream away from the surface to act like a jet that pushes the asteroid into a different orbit. However, advance testing is required to figure out how such explosions will actually affect the orbit of a known asteroid.

There are several unknowns associated with this plan that will have to be explored before we can rest secure that it might work. First we need to find out what a typical NEA is made of. Then we need to test it to see whether the technique works. I predict that after astronauts have visited several NEAs in the rendezvous missions now being planned, we'll be surprised that there are many different physical structures involved and that the deflection technique worked out for one type will not work for another. I'm not saying this because I have inside information on what asteroids look like up close, information that is actually available for some large objects (see chapter 3), but because nothing in the natural world is ever as simple as we expect it to be.

To be sure that we could push aside a doomsday asteroid we will have to test the technique, and to know, in advance, what the rogue object is made of. This may be asking too much.

A standoff explosion produces gamma rays, X-rays, and neutrons, all of which will penetrate the asteroid surface. At the United Nations conference on NEOs in April 1995, I asked what we could expect if a nuclear device were set off next to an asteroid but failed to deflect it. Would we be struck by a radioactive mass instead of harmless stones, rocks, and ice? The answer surprised me. The speaker said he preferred that I not ask, because it was a subject that would discussed at a meeting of "experts" a few months hence. I interpreted this extraordinary statement to mean that it would pose a very real, additional problem.

Instead of a standoff explosion, the nuclear device could be detonated on the asteroid's surface. This would dig a hole and the escaping gases would act like a jet to push the asteroid aside. Or the spacecraft could be flown headlong into the oncoming asteroid so that it buries itself deep inside before it explodes. But an explosion inside the object risks shattering it.

In order to make legitimate plans for asteroid interception and deflection, many more data are needed about the sorts of objects involved, and the orbits they occupy. Therefore, the first step to assure our long-term survival in the face of the NEA threat is to carry out a census to discover what's out there, how many of different sizes, what their orbits are, and what they're made of. Finding what's out there means constantly photographing the sky with suitable telescopes and cameras and doing so continuously, around the world (chapter 16). I suspect that once we begin this program it might be the one scientific experiment that will never stop, because the population of NEAs is constantly evolving. Therefore monitoring, once begun, may have to continue forever, or at least until we are struck by one that brings the program to an end, together with most of our otherwise civilized activities.

The Spaceguard Workshop also explored other ideas for deflection. For example, kinetic energy pulverization by means of a series of explosive darts or spears could wear away the object, a trick that would have to be performed far from earth to allow the debris to scatter. Other proposals included tying a rocket to an asteroid, mass drivers (steam rockets or electromagnetic guns), solar sails, crack outgassing, and laser deflection. Crack outgassing would involve creating cracks in the surface (by small local explosions) through which volatile material, such as ice inside the asteroid or comet, would escape and act as jets. Most of these options appear unreasonably expensive, and in any case still require a great deal of research to determine whether they are feasible.

To me it seems almost like a cosmic coincidence that about the time we discover the NEA threat we also have the technology (rocketry and nuclear devices) that offers us the chance to do something about the threat. Of course, there are fascinating technological challenges that wait to be resolved before a spacecraft can be expected to fly alongside an asteroid as a first step in a deflection mission.

On the other hand, as Duncan Steel has neatly summarized in his recent book, the simplest way to deflect an asteroid may yet be to slam into it at perihelion with a large mass either to slow it down or to speed it up by a small amount so that it will miss the earth.

The asteroid interception report summed up what steps would need to be followed in the years to come. The current searches for asteroids should be funded at a few million dollars per year, to enhance their present capabilities. These systems should then be improved, which will require about $10 million per year. Next, visits to asteroids should be carried out with light-weight space-craft for $100 million per year. These would be used to perform experiments on the nature of such objects.

The spacecraft *Clementine* was supposed to make the first visit to an aster-oid in about 1996, but it went into a hopeless spin after leaving lunar orbit in 1994. Since 1983, interested astronomers have been working with NASA on the Near Earth Asteroid Rendezvous (NEAR) mission. It was launched in February 1996 to visit 433 Eros in January 1999. NEAR will survey that rocky body for a year and come as close as 24 kilometers. This will pave the way for further stud-ies of NEAs. Some planetary scientists are already calling for a mission to test a standoff explosion at an NEA to discover what is involved in making a small change in the asteroid's orbit.

Only after these steps are followed can a reliable defense system be defined and put into place. The cheap part of the program is finding where the objects are and monitoring their orbits. Carrying out rendezvous missions then quickly adds to the cost. These escalate dramatically when the design and construction of launch vehicles for nuclear devices are undertaken.

When should we start on the mission to protect earth? The Spaceguard report stated, "The estimated level of threat demands a response now." Not everyone agrees. There are those who regard some of the people behind the Spaceguard program as fanatics exploring fantastic, costly schemes and that the task of deflecting an asteroid ultimately will depend only on our ability to push it into a new orbit. But whatever one's point of view, the first step is to carry out a detailed census of what's out there. As was mentioned before, there may be 2,000 objects of 1-kilometer diameter or greater in earth-crossing orbits and only 150 of them have been found to date. With a suitable network of telescopes, 90 percent of the remainder could be detected in 20 years. Using only present tele-scopes, it will take 50 years at least.

In principle, we have the basic technologies to eliminate the threat. With present weapons, smaller asteroids less than 100 meters in diameter can be com-pletely shattered or destroyed. Which option is chosen depends on what they are made of: metallic objects are not amenable to destruction, whereas stony objects are.

The most difficult objects to deal with will be long-period comets on their first visit. These offer very little advance warning. Comet IRAS-Araki-Alcock was detected on April 25 in 1983 by the Infrared Astronomical Satellite and confirmation only came on May 3. Its closest approach to earth was May 11 at a distance of about 5 million kilometers. Lead times of weeks for long-period comets may be the rule rather than the exception. If, however, a large comet were to enter the solar system and break up, the fragments, which might yet pose the greatest hazard, could be studied for centuries before they threatened the earth.

The Spaceguard workshop concluded that a standoff nuclear explosive of 100 megatons to deflect an asteroid was the most likely and efficient relatively short-term option now facing us. If the onrushing NEA was too large to be deflected by this means, a surface blast could be used. Then we'd have to live with the consequences of being hit by fragments. At least there might then be a slightly greater likelihood that most of us would survive.

It has been proposed that several systems be permanently installed to remain on standby for launch. The nuclear device would be brought out of storage and armed only when needed, and that would be done under international supervision. Having a number of defensive rockets on standby would even allow for a test run before the final thrust was imparted to a rogue asteroid. The technical jargon for this is that they would "fire-for-effect" to see what reaction is produced, and then plan the final launch detonation. This assumes we would have the luxury of being so careful.

All this would be expensive, up to a billion dollars if developed from scratch, and much cheaper if existing technologies are harnessed. Two or three years lead time would apparently allow plans like this to be implemented.

Just as we begin to relax in the secure knowledge that experts are thinking about these crucial issues here comes the bad news. Carl Sagan of Cornell University and Steve Ostro of NASA's Jet Propulsion Laboratory have pointed out that, as is true with virtually every major technology that can be used for good, asteroid deflection schemes when implemented might also be redirected for harm. Asteroid deflection technology could be used to redirect an otherwise harmless object to strike the earth. In fact, there is a large inventory of objects out there that could be so redirected, far greater than the number that will ever threaten our planet. "Possible offensive use, through error or madness, of an asteroid deflection capability strongly cautions against premature technical development," they wrote in *Nature* in 1994.

In other words, this technology designed to protect us, as with so many weapons systems in the past, may turn out to be the most powerful weapon ever invented. A subtle nudge with a 10-megaton nuclear explosion could redirect an asteroid to crash into the planet at a predetermined location to produce devastation far greater than could have been produced with thousands of nuclear bombs. A well-timed impact could get rid of all your perceived enemies with none of the unpleasant radiation aftereffects that would girdle the earth after a nuclear war (unless, of course, the asteroid is made radioactive in the process of deflection by a nuclear device!).

Sagan and Ostro concluded that "Premature deployment of any asteroid orbit-modification capability, in the real world and in the light of well-established human frailty and fallibility, may introduce a new category of danger that dwarfs that posed by the objects themselves." Whatever decisions remain to be made about a Spaceguard program during the next decade, Sagan and Ostro's caution concerning what is called the "deflection dilemma" deserves to be factored in.

Any realistic defense mechanism has to be in place ahead of time, and establishing a Spaceguard program will surely prove controversial. There are many ways to approach the problem, and at this time most scientists and military minds involved in thinking about how to protect the earth from impact have more at stake than just an astronomical interest in NEAs. Duncan Steel provides a sobering look at just what went on behind the scene of the Spaceguard committee meetings: "To someone familiar with civilian scientific conferences, however, it was startling how little science was actually introduced. With several happy exceptions, the response of participants was not tackling the problem in hand: it was angling for new big-budget programs and the power that goes with them."

If a plan to deflect asteroids is put into place, after identifying the threatening objects it would still be necessary to determine the environmental and economic consequences on impact. A plan to implement an intercept mission would then be based on what is learned. A deflection mission would inevitably cost billions of dollars and could not be done rapidly. If we are going to be serious about this, we'll have to start investing in our future immediately. Until we begin to think along these lines, the planet remains dangerously vulnerable to devastating impact. As Steel points out, "at the moment, we are not only without insurance, we also are not able to escape the house."

The nature of the impact hazard is global in scope. The United States has the technical resources to begin this work, but perhaps it can no longer afford to do anything like this. Perhaps the ultimate tragedy will be that our fate is sealed because we overspent in the early days of the space age. Our demise may yet be determined by the near bankruptcy of the nation best equipped to deal with the threat.

During 1995 a great deal was published about the programs to avoid future impacts. David Morrison posted a great deal of information on the World Wide Web (http://ccf.arc.nasa.gov/sst/), including an interesting excerpt from *SPACE-CAST 2020*, a U.S. Air Force publication. It pointed out that "Although not a traditional 'enemy,' the asteroids are nonetheless a threat that the Department of Defense should evaluate and prepare to defend against." They added that "provision for defense of the planet, as far from the planet as possible, needs to begin."

We take this as a sign that the U.S. military might well become a key player in the search for NEAs and creating a defense system to shield civilization from destruction.

Also posted on the Web was a summary of the Morrison Report on the Detection of Near-Earth Objects in which collaboration with the U.S. Air Force is encouraged and in which the overall international nature of search and mitigation efforts is highlighted. The cost of immediate search programs would be a mere $4.5 million per year for five years, after which the near-earth object census as regards the most dangerous objects would be all but complete.

The Morrison Report contains many striking statements. For example, "Expeditions to NEOs are logical next steps in an evolutionary program of human exploration of the solar system. The NEOs are humanity's stepping stones to Mars and ultimately deeper space." The report takes the daring step of suggesting that "The recent discovery that we exist in an asteroid swarm [see Figure 16-1] has enormous long-term consequences, and its historical importance may someday be seen to rank with Columbus's discovery of the New World." I would add that it is the discoveries of the asteroid swarm, the iridium in the K/T layer, and the crater that marks where a killer object struck earth 65 million years ago that are forcing us to reconsider who we are, how we got here, and where we are going.

Finally, the following quotation from the report: "The United States could celebrate the dawn of the Third Millennium by declaring its intention to land humans on an asteroid to conduct intensive scientific exploration and to return them safely to earth in the first decade of next century." This plan, if implemented would, for me, represent the coming of age of the human species. Walking on the moon was dramatic, but walking on a near-earth asteroid during its flight about the sun and past the earth would signify that our species had come to recognize that we do not live in splendid isolation from space and that we need to understand near-earth objects, comets and asteroids, if we are to live with them for a very long time into the future.

18
OUR PLACE
IN SPACE

*I*N July 1994 a spectacular display of violence illuminated the planet Jupiter. Had something similar happened to earth, few of us would be left alive to think about such catastrophes. Fortunately for us, comet SL9 missed by a very large distance. However, the equivalent of such an event did occur 65 million years ago—this is in a past so remote that it is difficult for us to imagine that it will ever happen again. Yet, somewhere in the earth's future lurks another fiery cataclysm. When it happens it will be no more or less than another instance of mindless violence that has shaped our planet and created the context within which life emerged and evolved. Comet and asteroid collisions with earth are crucial to the origin and evolution of life, and future visitations hold the key to our collective, long-term destiny.

Today we confront the stark fact that we live on a planet that has been shaped by violent and catastrophic events, catastrophic from the point of view of species that might have been unfortunate to experience them, yet creative from the point of view of those that followed. This circumstance is still with us. This, to me, represents the extraordinary breakthrough in our understanding of life in the universe that resulted from the discovery of iridium in the 65-million-year-old K/T clay.

The explanation of the mystery of mass extinctions might not be as emotionally satisfying as we could have hoped, given that we have recognized our

Figure 18-1 Comet Roland photographed on April 27, 1957. This comet developed a peculiar "antitail," which pointed toward the sun. It turned out that this was in fact a sheet of dust that had spread away from the comet and, in this exposure, the sheet was oriented edge-on to give the illusion of this spike. (Courtesy Yerkes Observatory, University of Chicago)

vulnerability in the face of an awesome cosmos. We live on the surface of a planet that is forever vulnerable to cosmic impacts capable of affecting our lives. We reside here, now, because of the way the processes of physics and chemistry and biology have played themselves out over billions of years.

Natural laws have determined how stars live and die, how and where interstellar molecules form, how those molecules were brought to an incipient planet, and how that planet was formed, deeply scarred by comet bombardments, and how during its first few billion years ever more complex organisms struggled to survive so that, ultimately, a species emerged that had the capacity to think about these things. That species then began to ask questions related to its own existence. Through dogged perseverance, it discovered answers whose truth could be tested against the reality of existence. It is only because we have made so much headway in understanding the laws of nature that we have learned to build electrical power systems, automobiles, planes, computers, and climate controlled housing—in fact, every material aspect of modern civilization.

In recent years, the message of comet and asteroid impacts (Figures 18-1 and 18-2) has been read and we realize that threat of impact will not go away

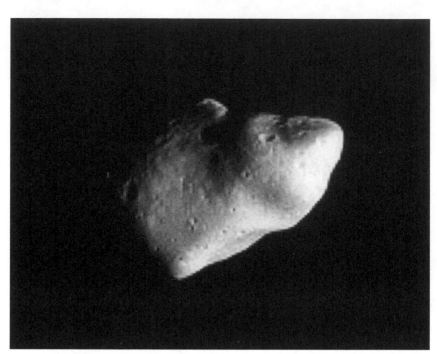

Figure 18-2 The asteroid 951 Gaspra as photographed by the *Galileo* spacecraft hurtling by at a distance of 16,000 kilometers on October 29, 1991. Collision with earth by an object this size (16 by 12 kilometers across) would devastate our planet and trigger another mass extinction. Gaspra, itself marked by craters, is currently a harmless member of the asteroid belt, but someday it may yet befall the human species to figure out how to deflect an object this size from a collision course. (Courtesy Jet Propulsion Laboratory/NASA)

because *Homo sapiens* emerged on earth to become cognizant of the danger. Yet this is not a new awareness; it has been lurking on the fringes of astronomical studies for centuries (see chapters 6 and 7), ever since comets began to be recognized for what they are, visitors from the depths of space. Once that was known, it was appreciated that they might occasionally slam into the earth with frightful consequences.

So here we are, basking in scientific knowledge. After a few million years of human evolution, which began after more than four billion years of life on earth, the mind of *Homo sapiens* is becoming aware of how it fits into the cosmic context. Some part of what we have learned about the ways of nature is truly terrifying, at least if we value our long-term survival as a species. We have learned that nature has awesome power to create, and to take away. If nothing else, this is the core message of scientific discovery. *Homo sapiens* recently evolved out of the mists of time to become conscious of this overriding fact. The message carried by the discovery of the threat of comet or asteroid impact is that we can no longer assume that civilization will go on forever. But once we become aware of this, what do we do next? Once we confront the fact that we could be wiped out in an instant of cosmic time, what do we do about it?

Our species may be passing through a phase experienced by a person who grows up and upon graduating from college must begin to function as an independent being. It seems as if we have spent a long time learning about the universe and suddenly we come face to face with the fact that we have been lucky in recent years. Our environment has, in general, seemed to be ideally suited to our survival. Apart from the atrocities we visit upon one another, person to person, tribe against tribe, nation against nation, and apart from the hazard of an occasional earthquake, hurricane, or flood, all seems well with the planet. But now we confront the fact that there exist forces in nature that will reduce all our human squabbles to insignificance.

Catastrophism forces us to confront apparent randomness—the unexpected coming out of the depths of night, as it were. No matter how we look at this, such a world view remains psychologically unacceptable for reasons that have nothing to do with scientific knowledge.

Childhood's end marks the time when we recognize that the decisions we make now will determine our fate, barring random catastrophes of course. In just this way, our species struggles toward mental maturity, and it has only just begun to confront an awesome perspective. No one promised us a rose garden; if we want one, it is up to us to plant it. This is what I see behind the message of the comets and asteroids. Will we do something about assuring our long-term survival? Based on what we have seen about past impacts, unless we take this issue seriously, now, it is unlikely that civilization, and probably our species, will have a long-term future lasting thousands of years.

To begin to take action means that we might have to transcend traditional thinking about how we came to be here, and where we are headed. To put this another way, to assure our long-term survival we may have to take charge of evolution in a conscious manner. For such a step there is no precedent in human history, or in the history of any species that ever roamed the earth. The point is that if evolution is driven by the occasional "random" impact that destroys a significant fraction of species alive at any time, taking steps to avoid a future collision is tantamount to seeking to control our destiny by avoiding what otherwise would have happened. Are we ready to join together to think about this? Can we even do anything to assure our survival? Should we bother?

Despite what we have learned about our future on earth, we show little likelihood of taking a real or potential threat seriously because of our belief in our importance as a species. Deep down, most people alive today probably believe they were placed here for a purpose, by some higher force, and that therefore they are protected from misadventure. Some people believe this so strongly that they are willing to kill anyone who disagrees with them. Isn't that strange.

There is an even more subtle reason why we are unlikely to take collective and significant action to assure the long-term survival of our species. It manifests as the psychological syndrome known as the "illusion of invulnerability." Individuals cannot believe that they will personally succumb in the next catastrophe. This syndrome is at play in those who live happily in earthquake zones, in floodplains, or on the sides of active volcanoes. The existence of the syndrome poses a paradox. If we are concerned about the long-term survival of civilization, we must overcome our genetic predisposition to deal only with the immediate future. Dealing with short-term issues is natural in all animals, and represents the practical way in which to survive from day to day. However, this predisposition is not conducive to assuring a long-term existence. Perhaps that is what is at issue. We have learned much about the natural universe in recent years, and the mind's eye has only just developed the ability to scan millions of years of time. Yet that seems to be no more than an intellectual exercise with little practical use. Perhaps the evolution of our species may yet depend on whether we can succeed in making very long term plans and carrying them out for the benefit of life on earth.

Scientific discovery has brought us to the point where we confront the awesome probability that collision with an earth-crossing object will bring an end to civilization. It is no longer a question of *whether* a massive impact will occur in the future; it is only a matter of *when*. Even if we think it will be a thousand years from now, the point of raising the issue is to ask ourselves what we plan to do about it. It may be time to think in terms of thousands of years into the future. I am assuming that we care that our species will be around for a long time, and that this question is worth thinking about.

This, then, is the dilemma raised by scientific discovery. Never before in the history of earth has a species existed that learned so much about itself in the context of what nature can do in the long term. That immediately implies that we are biologically unprepared to deal with this; we have no experience to guide us. We have only our intellect to point the way, and we've barely learned to trust that. We have to rely on the pictures we see with the mind's eye to begin to face the future in a deliberate fashion. If we value the continued survival of civilization, it is no longer sufficient to think only of tomorrow, or four years into the future, the limit in the United States between presidential elections.

So what shall we do with our newfound knowledge? On the one hand, we can decide to take steps to assure long-term survival, on a global scale. This will be tough. While human nature may work for individual survival, concerted action may be thwarted by social, political, economic, and religious persuasions that may prevent us from acting globally.

On the other hand, if we let nature take its course and accept our collective fate, civilization will sooner or later come to an end through, at the very least, comet or asteroid impact. (We'll ignore for now the danger that our species poses to itself.)

If we agree that the long-term survival of civilization is worth striving for, then making sure asteroids or comets do not strike earth is a quick way to focus our thinking. But when we confront the concept of long-term survival, we come face-to-face with the nature of human psychology. If we decide to avoid extinction by preventing a cosmic impact, shouldn't we also give more than passing thought to avoiding that same fate which may result from so many of the mindless activities with which technological societies threaten survival?

If we decide to do something about long-term survival, the starting point must be to recognize the underlying fact of our existence, that we find ourselves on a planet in a universe not specifically designed for our well-being. This is a capricious, violent place in which we live, and civilization will not continue just because we *wish* it so. We will have to take deliberate steps to make sure that it does go on, if that is our choice. Unless we take extraordinary and probably superhuman measures, civilization, and *Homo sapiens*, will go extinct. Of that there is no doubt. In the long run, the earth will be struck by a comet or asteroid large enough to do unto us what was done unto the dinosaurs. To pretend otherwise is to ignore the threat of the comets and asteroids.

We have recognized the existence of an overwhelming danger to our planet and to deal with it will require *conscious* and responsible decisions based on what we know about the nature of *reality*. But this seems to lead to a paradox. To assure collective survival consciously may require that we take charge of evolution. By that I mean we must make sure that certain events do not occur that would otherwise have happened, and vice versa. This will sound the death knell of the evolutionary process as it has acted in the past.

As soon as we begin to make conscious decisions related to our long-term survival, the nature of evolution will change. Early this century, the South African naturalist Eugene Marais made just this point in his book *The Soul of the Ape*. He argued that the inevitable consequence of the discovery of the *fact* of evolution brings about its end. Once a species becomes aware of this phenomenon, the phenomenon will itself be affected. Its something like the famous uncertainty principle in physics, which states that you can't know where an electron is without bouncing something off it to see, which means that you affect its motion. Thus the act of observation interferes with the phenomenon. You can never know precisely and simultaneously where the particle is and how it is moving. In the same manner, once we really observe the phenomenon of evolution, we affect it to the point of altering it.

Given what motivates humans to deal with the future, perhaps only a mini-disaster will help focus our attention, the wiping out of a few countries by a small impact event, say. That should help us to think and work together for a few years. But then we will inevitably slide back into the safe haven provided by the "illusion of invulnerability."

Another message of the cosmic threat of comet and asteroid impact is that the entire process of launching a civilization, from the time a species emerges that has the potential to become technological until it becomes aware of the fact that it has just done so, must take place between significant impact catastrophes. This evolution must take place during one of those moments of cosmic boredom referred to in chapter 1. In other words, we have been lucky to get this far. Civilization has emerged and evolved since the last major impacts, probably those which produced the floods of 7500 B.C. Even if you do not agree with the Tollmann interpretation of the meaning of sagas and legends, or the Clube and Napier variation on this same theme, civilization as a whole has had a fairly quiet time for many thousands of years.

In order to assure long-term survival, every action we take on a collective, and perhaps even a personal, level may have to be taken consciously and in full awareness of how it will impact our future. For this task we may, as yet, be lamentably ill-prepared, largely due to the psychological immaturity of our species.

The tragedy lurking within this perspective is that in order to survive we must let go of that which has sustained us up till now. Here the parallel between individual growth and that of civilization and our species is striking. As we grow up as individuals, we can no longer expect to be taken care of by benevolent parents. We must begin to fend for ourselves in just the way that the graduating senior must begin to look after himself or herself. Similarly, in the face of the comet and asteroid threat, we may have to learn to fend for ourselves. However, world events suggest that we are neither ready nor able to do so, unless we make superhuman efforts to confront the future in the context of the present, and do so in terms of what we *know* about the universe around us.

Our place in space is only likely to be assured if we begin to make conscious behavior decisions that will assure our long-term survival. Even as I make this statement, I am aware of a hidden assumption—that our species and life on earth are worth protecting. On two counts this assumption can be questioned. The first, which I have heard aired by some, is that the species is not worth preserving because the march of progress has been no more than an illusion. Deep down it appears that we are still savages at heart and will not be able to transcend our primitive instincts successfully, for example, to kill one another over beliefs, territory, or some material possession. A fresh start might therefore do our species a lot of good. The second is a more Zen-like point of view that leads to essentially the same conclusion. This perspective suggests we do nothing and let nature take its course. If a natural catastrophe were to wipe out the human species, only then would there be a chance for something "better" to emerge.

I agree with neither of these views. Instead, I am heartened by the increased attention now being given to serious plans to obtain a comprehensive census of NEAs and to the complementary plans to take steps to avoid impact. One can only hope that the momentum generated by these studies will touch a resonant chord in funding agencies.

It cannot be stressed enough that we know that asteroid or comet impacts have happened in the past. The evidence is everywhere. The impact craters on the moon and earth tell us so. We also know that the planet has seen many mass extinctions. Even Darwin knew this. He wrote, "Scarcely any paleontological discovery is more striking than the fact that the forms of life change almost simultaneously throughout the world." Now, 140 years later, the arguments about what triggered those changes has been settled, and the discussions can focus on details.

We also know that for every impact large enough to trigger an extinction event, there were dozens that only produced local catastrophes. Those will happen again, perhaps even soon. Some will be sufficiently devastating to push civilization to the brink without wiping out many species.

During the twentieth century at least 12 million people died in natural disasters such as floods, hurricanes, and earthquakes. Another 100 million were killed in wars. Should the Ultimate Impact occur, fatalities would reach five or six billion. This number paralyzes the mind.

There have been close calls in recent years. We almost got wiped out by a small asteroid in 1937, and again in 1989. To remind us of the danger, Jupiter was hit again and again in July 1994. As we ponder these things, scientists have begun to appreciate that comets (and their asteroid cousins), once essential for making a planet and creating the scene for life, have the potential to bring it to an end.

Seen from a cosmic perspective, the experience of being alive, and more significantly of being aware that we are alive, is surely not to be taken for granted. Let us enjoy every moment of our transient lives as we build up our combined awareness of what it means to live and survive on earth. If we then think we can take steps to assure the long-term survival of our species, and deem it worthwhile to do so, let us add to our knowledge base as we learn to live in the context of nature's awesome powers. Let us savor the moments of peace and tranquillity, for that is not really the way of nature.

From those early violent moments that accompanied the birth of our universe, through the chaos of starbirth, through the fiery cauldrons in the hearts of stars where the elements of life were cooked up, the stage was set for our emergence. Through the cataclysms of stars convulsing in their shattering death throes that created the heavier elements in supernovae, through the formative years of the solar system when collisions between planetesimals created the planets, the stage was built upon which we would, billions of years later, learn to comprehend these things. Through the continuing violence of internal upheavals that shaped the earth's surface, combined with cosmic bombardment that at first sterilized the planet and time and time again rendered countless species extinct, great changes were initiated that resulted in the emergence of *Homo sapiens*, a species that between major impacts has evolved to the point where it has become aware of these phenomena.

This, to me, is an awesome realization. We have incontrovertible proof that we live in a violent universe. We've known that for some time, but never before have we had to confront the fact that the potential for our extinction still resides within the solar system. We also appreciate that between moments of violence are long periods of calm, the length of time being proportional to the level of violence. There are long pauses between impacts that trigger mass extinctions, but there is much less time between impacts capable of shattering a nation. In the same way, there are long spells of good weather between the damaging storms. But when we know that storms are possible, we take steps to forecast their arrival and seek shelter when we see one coming. Will we do as much with the threat of comets and asteroids, many of which even now hurtle by our planet unseen?

During a debate on the cause of mass extinctions, we lose sight of the harsher truth that an impact will, sooner or later, wipe out civilization. That is the relevant thing for us to face, as a species. It doesn't really matter whether we understand all the details of past extinctions, or whether species went extinct rapidly or slowly. Seen from an astronomer's point of view, that is quibbling over details. It is how the awareness of our vulnerability moves us to change our perspective on life on this planet that may determine how long we survive into the future.

Deeply rooted in the discussions by every one of the authors who has dealt with the impact-extinction issue is the existence of a fundamental human bias, a bias that may be part of our genetic makeup which prevents us from considering the most staggering implication of all. We live in an apparently random universe that pays no attention to whether we are here to be aware of this fact. Violence, death, destruction, catastrophe, and cataclysm all visit planets at random. The violence of a comet impact is furthermore so vast that we cannot really imagine how our world will be altered after the event. Gone would be tranquil summer days with birds singing in the trees and light breezes rustling the leaves. Instead, there will be chaos on a planetwide scale.

This, then, is the nature of the world in which we live. We are perpetually poised on the edge of extinction. Now that we have learned this truth, what will our species do with its knowledge?

I am optimistic that during the next few decades we will put into place a planetary defense system. At the same time I hope that, as we begin to discuss this need in a broader forum, our combined awareness of who we are in the cosmic context will grow. If there is one thing the study of cosmic impact, the threat of comets and asteroids, has given me, it is a profounder sense of the nature of life on earth, our place in space.

BIBLIOGRAPHY

Abell, George. *Exploration of the Universe.* New York: Holt, Rhinehart & Winston, 1964.

Ager, Derek. *The New Catastrophism: The Importance of Rare Events in Geological History.* Cambridge: Cambridge University Press, 1993.

Alvarez, L. W., W. Alvarez, F. Asaro, & H. V. Michel. "Extraterrestrial Causes for the Cretaceous-Tertiary Extinction." *Science* 208 (1980): 1108.

Asher, D. J., S. V. M. Clube, & D. l. Steel. "Asteroids in the Taurid Complex." Monthly Notices of the Royal Astronomical Society 264 (1993): 93.

Asher, D., et. al. "Coherent Catastrophism." *Vistas in Astronomy* 38 (1994): 1.

Bailey, M. E., S. V. M. Clube, & W. M. Napier. *The Origin of Comets.* Oxford: Pergamon, 1990.

Baldwin, R. *The Face of the Moon.* Chicago: University of Chicago Press, 1949.

Beatty, J. Kelly, & David H. Levy. "Crashes to Ashes." *Sky and Telescope* (October 1995): 18.

Burke, John G. *Cosmic Debris: Meteorites in History.* Berkeley: University of California Press, 1986.

Chapman, C. R., & D. Morrison. Cosmic Catastrophes. New York: Plenum, 1989.

————. "Impacts on the Earth by Asteroids and Comets: Assessing the Hazard." *Nature* 367 (1994): 33.

Clube, S. V. M., ed. *Catastrophes and Evolution: Astronomical Foundations.* Cambridge: Cambridge University Press, 1989.

Clube, Victor, & Bill Napier. *The Cosmic Serpent.* New York: Universe Books, 1982.

————. *The Cosmic Winter.* Oxford: Basil Blackwell, 1990.

Dick, Thomas. *The Sidereal Heavens.* New York: Worthington Co., 1840.

Flammarion, Camille. *Popular Astronomy*. London: Chatto & Windus, 1894.

Gallant, Roy A. "Journey to Tunguska." *Sky & Telescope* June 1994): 38.

Gehrels, Tom. *On the Glassy Sea*. New York: AIP, 1988.

———, ed. *Asteroids*. Tucson: University of Arizona Press, 1979.

———. *Hazards due to Comets and Asteroids*. Tucson: University of Arizona Press. 1994.

Glen, William. *The Mass Extinction Debates: How Science Works in a Crisis*. Stanford, Calif.: Stanford University Press. 1994.

Goldsmith, D. *The Death Star and Other Theories of Mass Extinction*. Walker: New York, 1986.

———, ed. *Scientists Confront Velikovsky*. Ithaca, N.Y.: Cornell University Press, 1977.

Gregory, Sir Richard. *The Vault of the Heavens*. London: Methuen & Co., 1893.

Guillemin, Amédée. *The World of Comets*. London: Sampson Low, Marston, Searle & Rivinaton, 1877.

Herschell, Sir John. *A Treatise on Astronomy*. Philadelphia: Carey, Lea and Blanchard, 1935.

Hildebrand, Alan R. "The Cretaceous/Tertiary Boundary Impact (or The Dinosaurs Didn't Have a Chance)." *Journal of the Royal Astronomical Society of Canada* 87 (1993): 77.

Hildebrand, Alan R., & William V. Boynton. "Cretaceous Ground Zero." *Natural History* (June 1991): 46.

———. "Proximal Cretaceous-Tertiary Boundary Impact Deposits in the Caribbean." *Science* 248 (1990): 843.

Howe, Herbert. *Elements of Descriptive Astronomy*. New York: Silver, Burdett and Co., 1897.

Hsü, K. J. "Uniformitarianism vs. Catastrophism in the Extinction Debate." *The Mass Extinction Debates: How Science Works in a Crisis*. Stanford, Calif.: Stanford University Press (1994): 217.

Humboldt, Alexander von. *Cosmos: A Sketch of a Physical Description of the Universe*. New York: Harper and Brother, 1844.

Hut, Piet, et. al. "Comet Showers as a Cause of Mass Extinction." *Nature* 325 (1987): 118.

Kelly, Allan O., & Frank Dachille. *Target Earth*. Carlsbad: Target Earth, 1953.

Marais, Eugene. *The Soul of the Ape*. Harmondsworth: Penguin, 1969.

Melosh, H. J. *Impact Cratering: A Geological Process*. New York: Oxford University Press, 1989.

Milner, Thomas. *The Gallery of Nature: A Pictorial and Descriptive Tour through Creation*. London: W. & R. Chambers, 1860.

Mitchell, O. M. *The Planetary and Stellar Worlds*. New York: Phinney, Blakeman and Mason, 1860.

Morrison, David. Chair. *The Spaceguard Survey*. Report of the NASA International Near-

Earth-Object Detection Workshop. Pasadena, Calif.: Jet Propulsion Laboratory, January 1992.

Muller, Richard. *Nemesis*. New York: Weidenfeld and Nicolson, 1988.

Newcomb, Simon. *Popular Astronomy*. New York: Harper Brothers, 1879.

Ninenger, H. H. "Cataclysm and Evolution." *Popular Astronomy* 20 (1942): 270.

Norman, John, Neville Price, & Muo Chukwu-lke. "Astrons—Earth's Oldest Scars?" *New Scientist*, 24 March 1977, 689.

Norton, O. Richard. *Rocks from Space*. Missoula, Mont.: Mountain Press, 1994.

O'Keefe, J. A. *Tektites and Their Origins*. Amsterdam: Elsevier, 1976.

Proctor, Richard A., et. al. *Astronomy*. New York: Ginn & Co., 1926.

Rather, John D. G., Jurgen H. Rahe, and Gegory Canavan, co-chairs. *Summary Report of the Near-Earth-Object Interception Workshop*. Pasadena, Calif.: Jet Propulsion Laboratory, August 1992.

Raup, David. *Extinction: Bad Genes or Bad Luck*. New York: W. W. Norton, 1991.

Robin, E., et. al. "Evidence for a K/T Impact Event in the Pacific Ocean." *Nature* 63 (1993): 615.

Russell, H. N., R. S. Dugan, & J. Q. Stewart. *Astronomy*. New York: Ginn & Co., 1926.

Russell, H. N., *The Solar System and its Origin*. New York: Macmillan, 1935.

Sagan, Carl, & Ann Druyan. *Comet*. New York: Random House, 1985.

Sharpton, Virgil L. & Peter D. Ward, eds. *Global Catastrophes in Earth History: An Interdisciplinary Conference on Impacts, Volcanism, and Mass Mortality*. Geological Society of America Special Paper 247, 1990.

Shaw, Herbert. *Craters, Cosmos and Chronicles: A New Theory of Earth*. Stanford, Calif.: Stanford University Press, 1994.

Shoemaker, E. *Report of the Near-Earth Objects Survey Working Group*. On the World Wide Web, 1995.

Sigurdsson, Haraldur, et al. "Glass from the Cretaceous/Tertiary Boundary in Haiti." *Nature* 349 (1991): 482.

Smart, W. M. *The Sun, the Stars, and the Universe*. London: Longman, Green & , Co., 1928.

Smoluchowski, Roman, John N. Bahcall & Mildred S. Matthews. *The Galaxy and the Solar System*. Tucson: University of Arizona Press, 1986.

Steel, Duncan. *Rogue Asteroids and Doomsday Comets*. New York: John Wiley, 1995.

———. "Time-Dependance of the Near-Earth Object Population and Earth/Moon Cratering Rates." Paper given at a NEA Conference, 1994.

Steel, Duncan, & Peter Snow. "The Tapanui Region of New Zealand: A "Tunguska" of 800 years ago?" *Asteroids, Comets, Meteors*: 1991. Houston Lunar and Planetary Institute, 1992.

Steel, D. l., D. J. Asher, & S. V. M. Clube, "The structure and evolution of the Taurid Complex." *Monthly Notices of the Royal Astronomical Society* 251 (1991): 632.

Tollmann, Edith, & Alexander Tollmann. *De Zondvloed: Van mthye tot historische werk-lijkheid*. Baarn: Tirion, 1994.

———. The Flood Impact." *Mitt. Österr. Geol. Ges.* 84 (1992): 1.

Urey, H. C. "Cometary Collisions and Geological Periods. *Nature* 242 (1973): 32.

Verschuur, Gerrit L. *Cosmic Catastrophes*. Reading: Addison Wesley, 1978.

———. "This Target Earth." *Air & Space Smithsonian*. October/November 1991): 88.

———. *Hidden Attraction*. New York: Oxford University Press, 1994.

Velikovsky, Immanuel. *Worlds in Collision*. New York: Doubleday, 1950; reprint ed., Dell Books, 1967.

Watson, F. *Between the Planets*. Philadelphia: Blakiston Co., 1941.

Whiston, W. A. *A New Theory of Earth*. London: R. Roberts, 1696.

Yeomans, Donald K. *Comets*. New York: John Wiley and Sons, 1991.

NAME INDEX

SUBJECT INDEX